O wad some Pow'r the giftie gie us,
To see oursels as others see us.

— Robert Burns

Table of Contents

PART III *OUTLINING*

PART IV *RECORDING EQUIPMENT*

APPENDICES

From Deep Inside

Trapped in the cage of my own personality, I feel an urgent and personal need to communicate through music, which liberates me and stops the inner dialogue. In music I find the self of which I can be conscious only in performance, when consciousness expands. I work out my destiny onstage.

The destiny is mediated by notes. I must perform so they are transparent to the music. And the music must be transparent to the emotion. In performance, I have been conscious at times of neither notes nor music, but only of being a locus of activity, a streambed for a torrent of feeling.

I am greedy to spend all possible time like this. Everything of which I can normally be conscious must be mastered to clear the way for the magic which appears only in performance. From the hidden source of musical ideas, deepest feelings, magic visions beyond words, comes the call to voyage; and I must be ready.

Dear Reader,

Between the covers of this little book, as within the walls of a music studio, I offer for your consideration some thoughts on practicing and teaching, and also some very specific guidance on using tape recording as a tool in both activities.

I have several pictures of you before me as I write. In one, you are an amateur or professional musician of any instrument who wants to deepen your playing, but from within your own feelings, not those of a teacher or coach.

In another picture, you are a teacher aiming at a teacher's highest goal: that your students should not merely play better but know how to teach themselves.

In a third picture, you are a student, perhaps adult, perhaps quite young; perhaps advanced in your studies, perhaps only well begun. As a listener, you have discovered music's beauty and power; but have you developed as a performer the objective ear necessary to portray that beauty and communicate that power to others?

In the last picture, you are not a musician, not a music student, but simply one who enjoys eavesdropping on the "shop talk" of another field, hearing some amusing stories, and becoming acquainted perhaps with a slightly different world-view from your own.

Small as this book is, it was originally going to be even smaller, narrowly covering tape recording as a tool in practicing and teaching. But holding these pictures of you in mind, and trying to imagine your needs, I found I could not talk about using tape in practicing without saying what I think practicing is all about. The same went for teaching; so Parts I and II came to be broader than originally planned, and I hope more useful.

Whether we are professionals or complete beginners, much that goes wrong in our learning goes wrong at the very start of work on a new piece; so in Part III, I discuss a different approach to learning, one that lets us benefit from the virtues of taping right away.

I can hardly urge you to record yourself without giving you technical guidance in doing so, and I have tried to do this in a "low-anxiety" manner in Part IV. I discuss choosing and using the equipment for our everyday work, and also how to make high-quality tapes for audition or commercial release.

Throughout the book, I combine specific "what to do" with more general "why to do it"; the trees and the forest, you might say. I hope this approach makes you comfortable enough actually to try what is suggested here; for only in your hands and your work will my words find life.

Many ideas in this book may be new to you, but the very fact that you are reading these words shows that you are not afraid of new ideas. Just as we have a repertoire of pieces, we need a repertoire of ways of learning and a repertoire of ways of playing those pieces. We should embrace all the different approaches we can, secure that *we* will know what is best for ourselves, and that fruitful paths will automatically come to the fore.

I have called the chapters of this book "sessions," attempting to evoke the atmosphere of a coaching session. And just as what you learn with your coach or teacher comes alive only in your own practice, even more must any book rely on your independent work and sympathetic understanding. The give and take of coaching cannot be imitated in the one-way medium of a book, but I hope you will write to me with your comments.*

Until I hear from you, then, best wishes for your music-making!

James Boyk

*2135 Holmby Avenue, Los Angeles, CA 90025-5915
E-mail: boyk@caltech.edu

Part I

Teaching Ourselves

In practicing for performance, taping is a magic tool not exploited fully by most musicians. When they do use it, they tend simply to record and then plop down to listen, regarding the activity perhaps as a kind of licensed break from work, rather than an integral part of it.

I'd like you to see taping as not one activity but a whole candy store of them. Better yet, let's say an organic produce stand, because I believe you will find the activities not justy tasty and attractive, but musically nutritious.

Session 1

Energy in the Present Tense

We learn what we do, not what we say we're doing, nor think we're doing, nor intend to do later. If we want to perform, we must practice performing.

An immensely talented pianist friend of mine never practiced a performance. He would say, "Now I'm going to play the piece all the way through," and he would begin. He would play marvelously until he came to a place where he thought he had a problem. Without stopping, he would play that part again and again, looping around until he played it to his satisfaction. Then he would continue, but only until he came to the next such place, where the same thing happened. Listening to him practice made me want to scream.

If we want to perform, we must practice performing.

What he *did* was to practice playing the piece with loops around all those places; so that's what he *learned*: to play it with loops. What he needed to learn was to play it straight through; but as he never practiced this, he never learned it.

In performance, he didn't actually loop, of course; but you could hear where the loops had been, because the line broke, the thread was dropped, the pattern defocused.

You can avoid this problem by using the tape to perform for the audience that's always available: yourself. Here's how:

RECORD a whole piece, a movement, a section of a movement, an eight-bar phrase—it doesn't matter what, but play it straight through with performance commitment. As in any performance, use all your energy in the present tense. Is there none left for remembering what happened where? Fine; the tape will remember for you.

Next, TAKE A BREAK, anything from a walk around the room to a walk around the block. The goal of the break is to transform yourself from performer into listener, so don't go over the performance in your head. Think about chocolate cake, or basketball scores.

Now sit down comfortably to LISTEN, away from your instrument. Listen without the score, focusing on the big picture, and see what you think needs work. Jot down just a few notes on the important things, then go to the instrument and work systematically through your notes. Then listen again for the details, this time with score in hand. TAKE complete notes and WORK through them.

This gives you the idea, and may be enough for you to work with, in which case skip to Session 2.

Squinting Your Ears

Have you ever seen a graphic artist look at a paste-up to visualize how it will look in print? The paste-up has drawings and type and perhaps photos, each on its own piece of paper, all pasted onto thick cardboard. The edges of all these pieces won't show when the paste-up is photographed, but they are distracting to an inexperienced person. The graphic artist squints her eyes to blur the edges and see the design as it will look.

In listening to your tape, first "squint your ears" so you get the overall picture. Don't care about wrong notes and sloppy phrasing. Think about Time: Is it steady? Does the tempo work everywhere, or are there places where it's too fast or slow? Where is the place that *tells* you what the tempo should be? If there's more than one tempo, do the relations among the tempos make sense?

Think about Dynamics: Have you actually played loudest the place marked as loudest? How about the softest? Is the loudest place loud enough? Is the softest, soft enough? Is the average level right? Do the inflections speak and sing, or does the playing sound robotic?

"Top-Down" Work

You might listen to the tape two or three times to form a judgment about these things. In this work, you proceed like a painter, developing your interpretations "from the top down," designing the overall picture and then filling in the details. If instead you create nice details and glue them together, the proportions of the whole picture will be wrong.

> *Of what use is it to draw a hand beautifully*
> *if the hand is in the wrong place?*
>
> — Harry Carmean
> (Drawing Master, Art Center College of Design)

Of what use is it to draw a hand beautifully if the hand is in the wrong place?

To help you focus on the big picture, *de*focus the details. Don't take notes the first time you listen. Perhaps even lie down and close your eyes. Don't go to sleep, but a dreamy state helps.

Another thing which helps you to perceive the big structure of the piece and your performance is to play the tape back at double speed. If you are using an "open-reel" deck, record at 3¾ inches per second and play back the tape at 7½, or record at 7½ and play at 15.

One Thing At a Time

On your second listening for the big issues, jot down some notes to yourself, then work through them at the instrument. If the tempo wasn't steady throughout, practice playing the whole piece with steady tempo. If the dynamic shaping was wrong, practice that.

But when you do the one, don't worry about the other. When you pay attention to tempo, do some notes go wrong? Fine! Do the dynamics go out the window? Fine! You don't care; you're focusing on the one thing.

Obviously you have to be able to do it all at once. At some point, the "one thing" you will be working on is "doing it all." That will come of itself, when you have more experience with the particular piece and more with this way of working. For the moment, keep things simple!

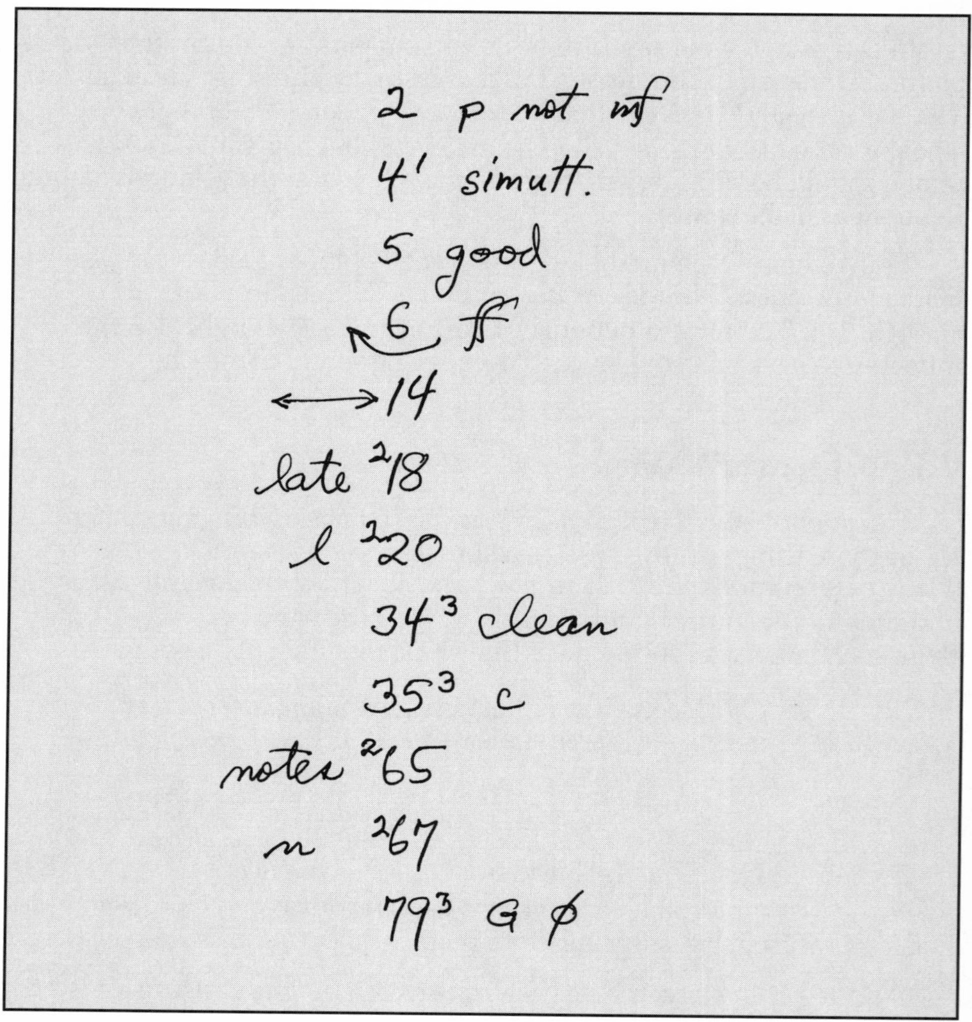

Illustration A. The author's shorthand for taking notes while listening to a tape. The idea is to write down just enough to jog the memory. The big number in each notation is the measure number. Comments to the right or left of it refer respectively to the right or left hand in that bar. Small numbers slightly above the bar number, if present, refer to beats in the measure.

The list shown means the following: In bar 2, right hand should play *piano*, not *mezzo-forte*. Bar 4, right hand, first beat: be sure chord notes are simultaneous. Bar 5 is good. Bar 6: play *fortissimo* (the arrow shows this applies to both hands). Bar 14, left hand: stretch the rhythm. Bars 18 and 20, left hand: the second beat is late. Bars 34 and 35, right hand: play the third beat cleanly. Bars 65 and 67, left hand: the second beats have wrong notes. Bar 79, right hand: the G on the third beat didn't sound.

Close-Up Listening

After you have worked through the large-scale aspects, you are ready to listen to the same tape again, this time focusing on the details.

It's important to mark down *everything*, for two reasons. If you notice the same things wrong the second or third time you record a piece, it alerts you that they are what my teacher Gregory Tucker called *errors*—things that need specific fixing—as opposed to random *mistakes*, which don't recur and are more telltale of carelessness.

We pay attention to everything for another reason, also, which is well expressed by the Talmud's advice:

> *To understand the invisible,*
> *look closely at the visible.*

To understand the invisible, look closely at the visible.

It might take two or three listenings to the tape to mark every single thing you hear. That's fine!

DOUBTER: "You mean I'm not actually practicing for half the time?"

Well, yes, you're practicing; you're just not playing the instrument. You are training your ear to be objective.

We musicians sometimes seem to feel like Antæus in mythology, who got strength from contact with the Earth. We get strength from contact with our instruments, and we don't want to let go! But it's the objective ear that makes music, and it's the ear we must train. This requires us to let go with our hands!

Maniacal Grin

Beware! Even when we let go, our muscles still play the piece again, covertly. If we're not careful, we will hear our muscles instead of our ears.

To avoid this, distract your muscles, including the important ones in the neck. Grin maniacally, float your head around, and flop your hands in the air.

I always feel like such an idiot doing this that I start laughing for real, which is very good for objectivity. It helps us be like Beethoven, of whom Tovey said, "He never hesitates to test his sublimest conception in the light of common day." We must never hesitate either; and laughter *is* the light of common day.

The hardest thing in the world is to be objective while involved. Performance involves us physically, mentally, emotionally and spiritually, so hearing ourselves objectively is four times difficult for us; and we won't attain it with blinkers and ear-muffs, but by opening our eyes and ears in every way possible. Laughter helps.

When you have your detailed written notes, work through them, checking them off as you go. If you marked a wrong note, play it right *once* (in context, of course), mark it off, and go on to the next thing in your list. Do it just once because that's all you need to call it to the attention of your unconscious; and because when you have done it right, doing it again risks doing it wrong. Never stop having just done something wrong; always "stop on right."

Taking notes and working through them sounds simple, but you will be surprised at how easy it is to digress. Keep yourself at it, though, and you will learn how well it works, and how focused it makes your practice. It gives you goals both short- and long-term; and your interaction with yourself through the taping process makes for intense work, just as does your give-and-take with your teacher in lessons. Time flies when you work this way; and when you finish, you'll be tired!

The closer you can approximate a true performance, the more useful it will be. Perform for others and tape your performance, even if it's just 10 minutes of playing for whoever is nearby. If you have a concert coming up, invite friends once a week for several weeks beforehand, and tape each performance. Spend the next practice session listening to the tape of one of the pieces and working through it. The next day, do another piece; and so on through the week and the entire program. By the next week's performance, you will be ready to learn from another taping.

Session 2

Dancing

Music-making is dancing which provides its own accompaniment. The dance may be subtle, but it's always there in the great performers. Conversely, I have seen one famous artist whose head starts operating separately from his body at climaxes. His head nods on the beat, but by the time the pulse reaches his hands, it is late. I have often wished I could give him just one lesson.

Musical time is fundamental to expression; and the particular aspect of time we call *meter*—the repeating pattern of strong and weak beats—is not so much a part of the expression as it is the very conduit through which the expression flows. Crimp the meter and you pinch off the flow of feeling.

Music-making is dancing which provides its own accompaniment.

Playing must be metric, but this does not mean metronomic. The metronome beats at equal intervals, which is sometimes appropriate, sometimes not. Even if equal intervals are wanted, using a metronome is still wrong, because meter must be generated within us, not come from without.

Many of the faithful apparently believe that a ticking metronome erases all sins and blesses all notes played in its presence; and we have all seen metronomes presiding over playing that is blithely out of time.

The metronome is anti-musical! Use instead the natural meter generator you always have access to: your body.

RECORD anything from a page to a whole piece.

DANCE to the playback. By dancing I mean any whole-body movement. Walking is fine; "conducting" and toe-tapping are not.

As you dance, NOTICE where your dancing body wants the beats to occur and where they actually occur on your tape. If the two disagree, the body is right; so PERFORM the passage again and dance it again. *Perform* it again, not *play* it again. You are teaching yourself to perform, and you do that by performing.

As you review a few passages this way, notice what kinds of errors you tend to make. Do the downbeats tend to be late? Do you accelerate when there's a crescendo, and ritard on diminuendos; or the reverse? Do you make gaps of articulation or pulse where one big section of the music meets another, or at moments of intense feeling? Such gaps are common "expressive" devices which kill true expressivity by stopping the dance. Even the tiniest gap cripples the musical line and emotional continuity.

Have you ever seen a room with bookshelves being built in? When they are all built and painted, they still don't seem integrated with the room. Then the carpenter puts caulking along the line where the bookcases meet the walls. Presto! Now they look built in. In other words, that thin line you didn't even notice killed the integration. The same happens with tiny gaps in performance.

> *The painting must have no holes*
> *through which the emotion can escape.*
>
> — Cézanne, according to Gasquet

In using this session's ideas over a period of time, you will be surprised at how your idiosyncracies recur in piece after piece. After some experience, they will lessen, or at least you will become aware of them more easily; but if you are like me, you will always have to wrestle with certain ones. I take this as a sign that the way we handle time is a deep aspect of our individuality.

The painting must have no holes through which the emotion can escape.

Of course we don't *want* to alter our individuality, just to eliminate those "tics" that constrain our expression. It's much easier to prevent them early on than to eliminate them once they're built in, so use dancing from the beginning of study. (For more on this see Part III.)

Originally I felt that "danceable" meant a listener should be able to dance to the playing on first hearing; the performance would tell him everything about when the next beat was coming. That idea was fruitful but not flexible enough. Now I think the playing should be "choreographable"; that is, one could dance to it naturally and expressively, but not necessarily on first hearing.

After dancing comes singing, the subject of our next session.

Session 3

Singing

Some may be naturally sensitive to music without necessarily being musically expressive. Yet everyone speaks his or her native language expressively. This opens the door to beautifully inflected playing via singing, the heightened sister of speech.

Some years ago I was asked to give a week of clinics in Canada for a hundred piano teachers from all over North America, who brought problems from their own practicing.

One teacher was browbeating herself over her difficulty with a lateral jump. The piece she was working on called for her right hand to play one octave followed immediately by another one higher on the keyboard.

Quick lateral shifts *can* be a problem in playing the piano; and those for whom they are a problem never seem to get taught the right way to do them. Telling "the right way to do it" makes me nervous, though, because people who think they *really know* how to play are almost guaranteed to be wrong. So I first took the purely musical approach, which *can't* be wrong. (It might be fruitless in a given case, but it can't hurt.)

Everyone speaks his or her native language expressively. This opens the door to beautiful playing via singing, the heightened relative of speech.

"Can you sing the interval?" I asked her. She had never tried, but gamely did so then and there. It was pretty bad; but after five minutes' work, she could reliably sing it.

Then I asked her to sing the interval several times in a row in a repetitive rhythm, ending by *playing* it without breaking the rhythm. She sang it perfectly, and then played it perfectly.

"So where is the technical problem?" I asked, palms up. "I guess there isn't one," she answered.

(Throughout this week of teaching, I noticed that my students seemed to feel deprived when a technical problem vanished painlessly. I regard this as highly significant, and we will come back to it later.)

When you, Reader, have a difficult sequence of notes, before you do anything else, make sure you can sing the passage not just comfortably but with verve. It is amazing how often this dissolves the problem.

If we can't sing it, of course we can't play it. And when we can sing it, difficulties vanish.

We can also use singing to make our playing emotionally more expressive.

PERFORM a phrase twice in the same recording, first singing it, then playing it.

LISTEN and compare.

Is your singing more expressive than your playing? If so, is it because of the way you handle Time? Is your singing more dance-y? Less clunky in the meter? Is the rhythm more fun, more precise?

Or is it your handling of Dynamics? Is the range of dynamics greater? Is the average loudness more appropriate? Are the inflections more convincing?

Or is it in some other area altogether? Your experience and sensitivity will suggest many other possibilities.

If your singing isn't expressive, if it doesn't move you, work on it first! Imagine singing the passage to three different people who evoke strong, but different, feelings in you. Or make up words which express the feeling; they needn't be coherent or even rational.

Words are useful for phrase-shaping also. It took me a long time to realize that the first four bars of Mozart's Sonata in F, K.332, didn't go to the middle or the end, but rather began with a pulse of energy which ran out over the course of the phrase. When I finally figured it out, I made up words which almost guarantee correct shaping of the first phrases.

Illustration B. *The author's words to the opening of Mozart's Sonata in F, K.332.*

Session 4

Spot-Checking

Usually a single movement has a single tempo, or speed (hence the name *movement*). In some movements, however, the tempo may change and then return to the original. How do we make sure the tempo really is the same when it comes back?

SET the tape counter to zero and RECORD the whole movement.

PLAY the tape and, with score in hand, LIST the counter reading at the beginning of every section whose tempo should be the same as the opening.

REWIND the tape and PLAY those places. Let the first one go on just long enough to get the tempo going in your body. Keep it going in your body while you fast-forward the tape to the next place, and play *it* just long enough to see if *its* tempo matches.

CONTINUE in this way through the whole movement. MARK on your list the places whose tempo was wrong, F for too fast, S for too slow. (It's easy to forget if you don't.)

RECORD the whole movement again, thinking about nothing but getting the tempos exactly right.

If this is too difficult, try the following:

In one continuous tape, RECORD all the places that should be at the same tempo. Play them one after the other, doing say eight bars of each.

PLAY this recording and CHECK by dancing. Do not be surprised if the tempos are different!

Why do I suggest listening just long enough "to get the tempo going in your body" or "to see if the tempo matches?" Because going on longer might reveal that the tempo *within* any given place is not stable, and that's not what we are working on. (If that is a problem, try the approach in Session 2.)

Scaling Tempos

Of course you can spot-check when you want different tempos at various places, also. I call this "scaling" the tempos because there is usually one place which either cannot be slower (because the dance would die) or cannot be faster (because it would have the wrong character or we simply could not play it). The tempo of this place becomes the reference to which all other tempos are scaled.

Dynamics

Spot-check dynamics, too. Compare all the places which are supposed to be *forte*, for example. Or, in one continuous recording, play a few bars of the beginning and then a few bars at each place where the dynamic level changes.

Of course I don't mean that tempos or dynamics should be mechanically identical or mechanically graded. Perhaps you *want* the triumphant return of the opening material to be a little slower than the beginning; enjoying the triumph, so to speak. Spot-checking helps you set whatever relationship you want, exactly as you want it.

Session 5

Beginning a Performance

The magic moment, when the room and all in it are encompassed by a deep and living silence. Out of the silence, the feeling of the music grows in you, becoming sound in your inner ear. When the music permeates you so its every aspect will be fully realized from the first sound the audience hears—Begin!

In assurance and stability, the beginning of the performance should be not like something starting, but like something which has gone on for some time and is only now being heard. To achieve this, start before you begin playing; but silently, of course. Here is a way to practice this.

START recording.

Be ready to play, but don't play yet. Instead, SING the first phrase of the piece, DANCING it in your body so far as your instrument allows. When you come to the end of the phrase, *without breaking the continuity*, begin playing the piece from the beginning (repeating the phrase you just sang). Play for two or three phrases.

LISTEN to the tape and see if this makes the beginning more focused.

Now make another tape. This time sing the piece more quietly, but make the dance as vivid as before, with strong metric feeling and rhythmic "point." Again, begin playing without breaking the continuity of the meter.

Finally, make another tape. Do not sing out loud, but *hear* the song in your head, dancing it just as vividly as ever. Begin playing as before.

Your playing for the last tape should have just as much stability as for the first. If it does not, I suggest doing a series like this every day for a few days. Each series will take just five or ten minutes.

The magic moment, when the room and all in it are encompassed by a deep and living silence.

PART II

TEACHING OTHERS

All teaching is learning; all learning is teaching. We teach others or we teach ourselves; we are taught by ourselves or taught by others. In learning to teach others more effectively and humanely, perhaps we will learn the same about teaching ourselves.

Session 6

Learning Without Anesthetics

A Child Learns

When a child learns to turn a doorknob and open a door, no one says to her, "Now Mary, flex your fingers, supinate your forearm, and extend your elbow." Probably no one says anything. The child sees people opening doors, so the goal is well-defined. The sheer power of wanting the goal teaches her. Once she has done it two or three times, the skill is hers. We call "gifted" a child who is specially quick, but every child learns the skill.

Does anyone think we could teach the child even such a simple skill *better* than she teaches herself? Playing an instrument is much more difficult, yet we think we can teach *it* to people better than they could teach themselves.

I think correct teaching is based on a realization that humans are goal-oriented beings who teach themselves well when the goal is vivid. Our job as teachers is *to make the goal vivid* and *to give instant reinforcement* of steps in the right direction.

The Goal is Sound

In music, the goal is sound; so training the ear is primary. The best teacher of any ear is that ear itself. Taping helps the ear to hear for itself, and thus speeds the process. Then the teacher can function as a guide familiar with the territory, pointing out shortcuts, scenic beauties, and the paths to the peaks.

Humans teach themselves well when the goal is vivid.

But the idea of teaching as a process performed on the student by the teacher—I don't believe that happens, or can happen, in music or any other field. A student is not a product, and learning is not a "feature" which can be added to the product.

Certainly there are students who want to be operated on. "I don't want to know about it," they seem to say to the surgeon-teacher. "Just knock me out and make me a good player."

I don't know why they bother. To me, the fun is to find my own path. If it turns out to be already well-trodden by others, that's fine, and amusing, too: All that work just to learn for myself what others already knew! If it turns out I am the first one on the path, that's also fine, and exciting.

The Power of Wanting

When I heard the great jazz pianist Thelonius Monk in a crowded Hollywood club, I arrived so late that there were no regular seats left and I was seated on stage with him, about where piano teachers sit relative to their students, close enough to touch him and to see how he was playing. Of course this was just what I would have asked for!

Monk's use of his body looked so strange to me at first that I thought, "If I saw a silent film of this, I would swear he couldn't play." I would have been wrong. He played wonderfully, every note to the point.

I could almost *see* his internal ear hearing the music and guiding his body. The music was full of surprises and delights, one door after another opening into treasure rooms. Monk *wanted* those doors opened, I thought, and his wanting was so vivid and precise that his body achieved the goal.

Session 7

Teaching

> ***"Play it any way you want–as long as it's wonderful."***

My lesson with Gregory Tucker was in progress in the big ground-floor room of the old mansion that housed the Longy School, in Cambridge, Massachusetts. I loved that worn and handsome room, and often practiced there in the evenings. More than a quarter-century later, I can still see in memory its generous, irregular shape with big bay windows, ratty oriental carpets on the hardwood floor, the high fireplace which had not seen a fire for many years, and the two Steinways at which our lesson was progressing.

This lesson was not progressing. I loved my teacher, but on this day I resisted his every suggestion and argued about every idea he offered. Finally he said, "Look, Boyk, I'm not saying play it my way. Play it any way you want—as long as it's *wonderful.*"

Power Struggles

That brought me out of it with a start. Mr. Tucker had short-circuited a power struggle that was wasting energy and time. Every teacher needs this ability (and the ability to recognize when it's needed.)

Here is an approach which helps avoid power struggles altogether:

RECORD the student's playing.

LISTEN together, and ask the student for a critique.

CONFIRM the student's perceptions of her own playing, and PRAISE her for them.

By this simple process, you take an occasion for criticism, namely, the critiquing of the student's playing, and turn it into an occasion for praise of her perceptions. Instead of potential opponents, you are partners.

In addition, you teach the student to depend on her own judgment, which is after all the goal of teaching. Then you can fulfill your proper role: supplying whatever subtleties of perception, knowledge of music-making, and instrumental mastery the student does *not* yet command.

You may ask, "But what if the student's perceptions are wrong? How am I then to give confirmation and praise?"

In my experience, the student's perceptions won't be wrong, or only rarely. And if some are wrong, some will be right; and you can praise *those*. If necessary, go back to basics. Dance to the tape together, and have the student do the singing exercise.

An Archive of Student Playing

Another simple and valuable practice is to set aside one tape for each student. Every six months or so, record the student's performance of one or two complete pieces, adding these after the previous recordings on the same tape, with a recorded announcement of the date. This can be done in the lesson, with you as the only audience; but of course you can also record performances.

As the student listens to any one of these recordings over time, she will gradually be able to listen without covertly playing the piece (see Session 1, above), and will thus become more objective about that particular performance.

Have her listen to the same tape and write down her judgment of it, say every three months; and see how long it takes before the judgment stops changing! It can take as much as two years for the judgment of inexperienced players to stabilize; two years, that is, before the student hears her original performance objectively. In some ways, it is only then that learning from that performance can begin. (Meanwhile, her study of other pieces has continued, of course.)

Here then is one concrete measure of the task of educating a musician: To reduce the two-year delay in objectivity to zero, so she is aware of what she does as she does it. I name a two-year period because that's what it was for me when I began. For a more gifted person, it might be zero to begin with.

To Hear Ourselves as Others Hear Us

RECORD a lesson you give, and a day or so later, LISTEN to the tape of yourself teaching.

We who teach should listen to these tapes only when we feel strong, because we will hear ourselves offering contradictory statements and unclear explanations, omitting explanations, and not responding to what the student actually asked. We will take too much time on one point while skipping others equally important. We will play examples that don't show what we say they show. We will be impatient, rushing students into attempting what they have not understood. Or we will sound patient merely because we are half asleep!

Many students don't feel they're being taught unless the process is more or less painful. Our next session tells more about this.

Session 8

Of Conflict, Exhortation, and the Student-Teacher Tango

A master class in the home of an older pianist of European background, internationally known. The comfortable living room is dominated by two big Steinways. The listeners, mostly older ladies, pay to attend. The performer is a student of the famous one, a young woman whom he has chosen to play Chopin's G Minor Ballade. She also pays.

She does not play well. When she finishes, all heads swivel to Der Meister, who sits in a wing chair in the corner. Long silence. A deep sigh. Then—he speaks!

"Ach, it is so ungrateful a task to destroy that which has been built up!"

He goes to the piano and sits down on the bench, which the student vacates; and while she listens, standing awkwardly beside the instrument at the bass end of the keyboard, he gives a beautiful performance of the piece, following it with a charming informal talk.

That is, it would be charming had the whole sequence not made clear that the bad playing by the student—the student he chose, dear Reader—was merely setting the stage for the good playing by the teacher. In the hour and a half of the class, he says not one word, shows not one way to play a passage, does nothing that helps his student play better. Except by the example of his own playing, of course; but if the student were able to profit from that, she would have played better to begin with.

Every Teacher a Master In His Own Studio

My friend Lincoln Mayorga, at age 15 or so, played at a music festival master class of another famous teacher. The teacher had rented for the season a gracious house whose living-room window gave on a beautiful old tree just outside and, beyond it, a view of the high-mountain summer foliage. In the living room were the teacher's two pianos, and here he conducted his master classes as a series of private lessons with audience, the audience being his other students.

As Lincoln tells it, before each class, the teacher's wife would read a list of the students whose payments were late—"Stevens owes us for two lessons; Steinmetz for one; Goodale for one..." (these names are fictitious)—while the teacher feigned impatience and indifference:

"Please, Rose, we have work to do. Really, Rose, can't this wait? Rose, please, we must begin now."

Lincoln played a big piece by Brahms, and was a little unnerved to have total silence greet the end of the work.

"Lincoln," the teacher finally says, "Lincoln..." He relapses into silence.

"Lincoln..." Silence.

He rouses himself to consecutive speech.

"Lincoln. Look out the window, Lincoln, and tell me what you see."

Lincoln looks out the window and says, "I see a tree, Maestro. I see leaves, green."

"Of course you see a tree, Lincoln. We all see a tree. But what do you...*see?*"

Lincoln looks out the window, hesitates, and says, "Beauty! I see natural beauty, Maestro, the beauty of the world."

"Of course you see the beauty of the world, Lincoln. We all see the beauty of the world. But don't you understand, Lincoln? What...Do... You...*SEE?*"

Lincoln turns his head toward the window, but he is not looking out; he is racking his brain. As though from a distance he hears the teacher's voice:

"Lincoln! Lincoln, do you not see...*ST-RUC-TURE?*"

The Limits of Exhortation

In the exhortative method, the teacher simply *demands* better performance: "Play more beautifully!!" Exhortation is quick, because no time is wasted on explanations. When coupled with imagery—especially if the teacher can play an example herself—it can efficiently give a picture of what is wanted: "Play it like water dropping!" "Like molasses!" "Sexier!" "As though in a graveyard at midnight, you walk into a cobweb." "Like a bird on the wing!"

Exhortation, imagery and example are wonderful–until they fail.

Exhortation, imagery and example are indispensable; and they are wonderful—until they fail.

Then the teacher must have the insight to see what the student needs, the knowledge to supply the answer, and the ability to explain it so the particular student can understand it.

Otherwise, we have a teacher who in this particular case cannot teach, exhorting a student who in this particular case cannot do. This is a perfect setup for sadistic teachers and masochistic students.

Le Geste

I was in the middle of a six-hour lesson with Aube Tzerko. Another of his students had joined us, and was sitting on the little three-legged stool behind and between Mr. Tzerko and me where we sat at the two Steinways.

Mr. Tzerko left the room for a moment, saying we would work on the Opus 111 when he returned. This is Beethoven's last piano sonata, a sublime piece regarded as almost sacred by many musicians, of whom I am one. It ends with a vision of Heaven, and begins in the most unsettled and tragic human feelings. Even the first notes are turbulent, a commanding gesture of two octaves in sequence, played quickly by the left hand alone.

When Mr. Tzerko had left the room, the other student leaned forward and asked me, "Do you *play* the 111?"

"Well," I said, trying to be modest, "I'm working on it, yes."

"No, but do you actually *play* it?"

"Well, I haven't performed it; but if you mean do I play the notes, yes, I do."

I couldn't figure out what the fellow meant. Then he told me his story. A Midwesterner, he had recently come to Los Angeles after two years of study with a famous teacher in Europe. He too had worked on the Opus 111, but he had not actually played the notes. Instead, his European teacher wanted to see the *gesture* with which he proposed to play the opening two octaves. "My boy, let me see *le geste*," she would say.

He was a little startled at first by this extremely "Zen" approach; but after all, he thought, she was a famous teacher and he was a nobody.

So he would draw himself up. He would think of the depth of meaning in every note, of the tremendous spiritual journey in the work, and of the gates of Heaven opening late in the second movement, 24 minutes away from the two notes he was now contemplating.

With complete focus, he would make *le geste*. "No, *no*! You ***still*** haven't got it," she would scream; and she would send him home to work some more.

In the lesson, no notes had been played.

A Frustrating Experience

The student was playing a piece that had a repeated two-note group (a descending slurred half-step) marked *ritardando*, that is, "getting slower." The music had no words, but for clarity, let's put the words "Love me" to each pair of notes.

The student played a conventional *ritard.*, in which successive notes got farther and farther apart. We might represent this as follows:

Love me, Love me, Love me, Love me

The teacher said, "No, don't do it like that. Do it like this!" and played a *ritard.* in which the spacing between the two notes of each group stayed the same while the intervals between groups got longer. It might be represented this way:

Love me, Love me, Love me, Love me, Love me

Thus exhorted, the student played the passage again exactly as before.

"No!" said the teacher. "Do it like this!" and played it again *his* way. The student played it *his* way. The teacher played it *his* way.

The student *did not hear* that the two ways were different, which in a college music major shows a lamentable—but common—lack of an analytic ear. The teacher heard they were different, but either did not realize this needed saying, or did not have the verbal skills to do so, either one a lamentable—but common—deficiency.

I was a guest and kept quiet. The miscommunication never was cleared up. After a few minutes, they moved on to something else.

Inner Conflict

From many years of experience as both student and teacher, I have learned that the psychological element dominates learning in music, as in everything. Motivation achieves; conflict murders achievement.

Motivation achieves. Conflict murders achievement.

The conflict is sometimes between teacher and student, but it can be entirely within the student, as seemed to be the case with one intermediate-level player who came to me. The only member of his family not a professional musician, he worked in a technical field to which he was a brilliantly creative contributor.

He brought one of Bach's so-called "little" preludes to his second lesson, and played it badly. Something about the playing puzzled me. It is hard to remember what was wrong, but I retain an impression that the playing seemed bad in the wrong way. The wrong things were wrong, so to speak.

In whatever way, it struck me as unlike anything I had heard before, and it led me to do something I have never done before or since. When he finished the piece, which was very short, I said, "Please play it again," and I kept saying this until he had played the piece seven times.

The second time was distinctly better, not just in technical details, but in the whole feeling of music being made.

"Please play it again." The third time was better yet.

By the sixth repetition the playing was really excellent: lively and fluent. I began to wonder if I should continue asking for repetitions, but for one last time I said, "Please play it again."

The seventh time was terrible, worse than the first!

"This is not a musical or pianistic problem," I said.

"Yeah, I know," he said; and that was our last lesson.

In my view, *if* doing music was important to him, he needed not a music teacher but a psychotherapist. As I saw it, his considerable musical gifts were turned back on themselves ("root-bound," they call it in plants) by internal conflicts about music. What was the nature of these conflicts—if conflicts there were—I can only speculate. For a multi-gifted child growing up in a family of musicians, perhaps there were substantial pressures in following the path towards music. Perhaps it was easier simply *not to be good* at music than to be good at it and *choose* not to do it.

For myself, I was sorry to stop teaching such a very interesting person; but at least we were not wasting our time and his money working on things that were not the problem.

John Henry

'Cause before I let that steam-drill beat me down,
I'll die with my hammer in my hand, Lord, Lord,
I'll die with my hammer in my hand.

— "John Henry," American folksong

John Henry was a martyr to machinery. Some musicians feel they must be martyrs to the music. A piece of music, they feel, is a test which mere human abilities, theirs at least, cannot pass; and ideally *they* should be some kind of machine. This is wrong. Music is of people, by people, and for people, not machines!

But these poor people seem to have a vested interest in proving themselves right by not playing well, and they often choose teachers who prove them right by demonstrating at every opportunity that they, the students, do not understand, cannot play, and are not worthy.

"It takes two to tango," and such teachers may accept such students because they need someone to whom they can feel superior. This is satisfying to the teacher's narcissism, but it means the teacher has a vested interest in the student's not improving, and it represents a dead end for the teacher's own growth.

The student in such a relationship may actually resist good teaching, unconsciously thinking along these lines: "I know I am not capable of understanding the real content of music. Therefore, since I understand what you are teaching me, it can't be substantive. Therefore you must not be a good teacher."

Gentle Reader, this happens!

Two Tangos

TEACHER: Play it *this* way. (*Describes verbally or plays example.*)

STUDENT (*plays*): Was that right?

We ought to put these lines in huge bold type, because this simple exchange, which is so common, is already going in a dangerous direction. What's potentially wrong becomes clearer as it continues.

TEACHER: No. Do it again.

STUDENT (*plays*): Was *that* right?

TEACHER: Well...it was better. Now *do* it that way every time.

STUDENT (*not at all clear about what he's supposed to be doing*): OK.

Here is a possible alternative, starting from the beginning:

TEACHER: Just as an experiment, try playing it *this* way. (*Describes verbally or plays example.*)

STUDENT (*plays*): Was that right?

TEACHER (*with a sympathetic smile at student's not relying on his own perceptions*): You tell *me*!

STUDENT: I don't know.

TEACHER: OK, play it again and listen to it as you play. (*Student plays.*) What do you think?

STUDENT (*who has listened this time*): I *think* it was what you asked for.

TEACHER (*confirming student's independent perceptions*): I think so too! Now what do you think of that way of playing it?

STUDENT: I don't like it.

TEACHER (*praising student's relying on own judgment*): Good. *Why* don't you like it?

...And so on.

This exchange takes more time, but the time is spent directly in developing the student's reliance on his own ear and judgment, which is, after all, the goal of teaching.

Though taping has nothing to do with my point here, it does have an important potential role. When the student says, "I *think* it was what you asked for," the teacher, instead of directly confirming that it was, can suggest they listen to the tape, which will allow the student to give his own confirmation. The teacher can then reinforce *that*. And when the student says, "I don't like it," if he has trouble explaining *why* he doesn't like it, listening to the tape may help him clarify it. In both, the tape allows greater objectivity and accuracy by separating the act of paying attention from the act of playing.

TEACHER: *Play it this way.*
STUDENT (plays): *Was that right?*

We're all worthy of music's blessings!

Exhortations to the Reader

Don't denigrate someone else to prove your worthiness to yourself! Don't use music to prove your unworthiness to yourself! We're all worthy of music's blessings!

Part III

Outlining

Just as our early education determines much of what we grow up to be, our mature musical interpretations are formed by our first experiences of the pieces. The moment of opening a new score is magic, never to be repeated!

To preserve the magic, practice must preserve a direct, not convoluted, relationship to performance. Too often, the meaning of a performance seems to be the practice that led up to it, and the performance itself becomes a recounting of the tortures of learning. This is backwards: *The meaning of the practice must be the performance!*

Twenty-five years ago my own practice was revolutionized by my discovery of Abby Whiteside's books, which I studied on my own and in a week of daily lessons with her pupil, the remarkable pianist and composer Robert Helps. In the following group of sessions, I introduce Whiteside's idea of "outlining," which lets us all approach new pieces as the most gifted players do. I do not simply refer you to Whiteside's books, because I want to put outlining into context with the approach I describe in this book.

I think Whiteside's ideas apply to every instrument. I confess, however, I have had little luck convincing non-pianists to try them. However, if you play an instrument other than piano, it will be easy for you to adapt the examples here to your own scores.

Skip this Part if you wish, but you will miss some good stories!

Session 9

The Road Less Traveled

Wouldn't it be wonderful if we could use taping from the beginning of studying a new piece? There is a way, not hard, but less traveled. Many people are put off by how different the musical scenery appears from this path, and they falter before they have properly tried it and found its pleasures. Those who succeed carry curiosity in their packs, and a sense of humor.

When I say leave out notes, I don't mean occasional ones. No, leave them out in bunches!

The secret is to leave out notes, and to use the resulting ease to give a fully involved performance from the beginning.

When I say leave out notes, I don't mean occasional ones, such as you have done in practicing ornaments, leaving out a turn or a trill. No, leave them out in bunches, so hardly any remain! Retain for your "outline" only those notes most essential for the meaning of the music. Succeeding outlines will gradually fill in the picture, until the last outline is the piece as written.

In the examples below (Illustration C), I have marked a first outline one might play for Bach's Prelude I from Book I of the Well-Tempered Clavier, and a possible second-level outline for Chopin's Prelude in G, Opus 28, No. 3. Even looking at these without playing, you will understand why, to the eyes of many, it feels odd to jump from one outline note to the next. It makes you a mountain goat, gamboling among the musical prominences.

This is where many falter. For myself, when I began outlining, I thought it was hilarious that after reading and playing so much music for so many years, I could be disconcerted simply by a new way of looking at a score. But after all, we know that good readers do grab notes, bars, and phrases in bunches visually, just as good readers of written language grab words and phrases in bunches.

I turned pages once in college for Prof. Luise Vosgerchian, a superb musician and phenomenal reader, as she sight-read something in class. I was startled when she bobbed her head for the page turn about two and a half lines before the end of the page!

I had been startled in the reverse way two years previously, when a high-school teacher read aloud to us. He labored through the lines like a mason laying down bricks, one ponderous word at a time.

Illustration C1. *Possible first outline for Bach's Prelude I from Book I of The Well-Tempered Clavier.*

Illustration C2 (below). *Possible second outline for Chopin's Prelude, Op. 28, No. 3.*

Good readers don't read like that, whether in language or music. It's not the speed that's wrong—that's just a symptom—it's the whole process. And that's why, when I came across outlining, I immediately felt it had something to offer me.

With just this introduction and the two musical examples to point the way, you may wish to follow the path yourself, experimenting with a new piece. (Don't use a piece you already know.) Or you may wish to read the next two sessions first.

Session 10

From the Beginning

If you have walked with me through Session 9, many questions may be slowing your steps. How does one decide which notes should be in the first outline? What notes should one add next? In short, what exactly is the right path to follow?

Let us lay these questions aside for one session, put down our packs, and step out on a promontory to get a view of the territory. When we see our destination in the distance, and the lie of the land between, we will not worry so much over the details of the maps.

Interpretive Infanticide

By the time she felt ready to study "the interpretation," it was lying dead on the practice room floor.

Why are so few performances moving? Or rather, why are so many performances moving only intermittently? The artist is sincere, and has worked hard. The technical equipment seems to be there, the notes are correct, the musical sensitivity is obvious. It is hard to put our finger on what's wrong or missing, yet we're not moved. "He's just not a great artist," we say, or "She had an off night."

Some nights *are* "off" nights; some artists *are* great. Years ago, my wife and I turned on the car radio on the way to meet her parents for dinner, and heard a man reading "The Rime of the Ancient Mariner." In high school, I had recited the opening section from memory; but not like this! This was so riveting that although we arrived outside the restaurant in plenty of time, we were late getting inside because we stayed in the car to hear the poem all the way to the end. Thus I learned of my own knowledge that Richard Burton was a great reader.

I am not suggesting that we can all read like that—at least I'm not suggesting it in this book. But we all speak our native language fluently and expressively, at least up to the natural expressiveness of our natures. Why not the same with music? What *is* wrong with those performances that don't move us?

Fundamental errors were built in from the beginning! By the time the artist "knew the notes," she had committed infanticide on her performance. By the time she felt ready to practice and study "the interpretation," the interpretation was lying dead on the practice room floor.

The performance is dead because *the dance of feeling carried in the dance of the body* was ignored while the notes were learned. The solution is to learn the dance *with* the notes.

Ideal Learning

Of course, some people don't seem to need a learning process. When Grieg brought the manuscript score of his piano concerto to Liszt, with the ink hardly dry, Liszt played it at sight in such a way that Grieg said, "No one will ever play my concerto more beautifully."

Unfortunately for me, I was not present on that occasion; but some years ago, I was present at a "Messiah" party, an informal holiday gathering to sing Handel's great oratorio. The accompaniment was supplied by one of the guests, Richard Grayson, who sight-read a "reduction" of the orchestra score (an arrangement for piano of what was written for orchestra). Watching over his shoulder, I was startled that he played so well, especially in one section, a fast and complicated fugue.

"I knew you were an amazing reader," I said, "but that's incredible."

"Oh, I wasn't reading it," he responded.

"You've studied it, then, or played it before?"

"No," he answered.

"Wait a minute. If you weren't reading it, and you haven't studied it or at least read it before—well, what other possibility is there?"

His answer floored me. "Oh, I wasn't reading it. I could never read that; it's too difficult. I was just *improvising along the lines I saw Handel following.*"

Many years before, as a teenager, I studied at Interlochen with Henry Harris, who always said that if you were ready to learn a piece, you could learn it fast. Conversely, if you couldn't learn it fast, you weren't ready to learn it. You might still work on it as a vehicle for developing your abilities, but you weren't doing it simply to expand your repertoire.

Some people don't seem to need a learning process.

I got so frustrated at being taken to task for being slow, that one week I resolved to learn some piece, however short, before the next week's lesson. I went to the music store and happened on a set of five very short pieces by Prokofiev, the "Sarcasms," and learned three of them that week.

At the next lesson, I said, "Mr. Harris, in view of your attitude toward my slow learning, I have brought you three Sarcasms." He grinned.

I played the pieces for him, and he said, "Very good, pal. I haven't seen these. May I sit down for a minute?" He sat down at the piano—and played all five at sight better than I played the three!

Moral Virtue

There is no moral virtue in practicing hard. The virtue is in playing beautifully. The ideal is that on first seeing a piece, we would immediately play it in a finished way, as these people did.

Opposite Answers

What prevents us from doing what Liszt or Henry Harris did? It's that a new piece confronts us with *too many things to think about in too little time.*

The usual solution is to use more time, that is, do "slow practice." But changing the tempo changes the feeling, so then we are not practicing the actual emotion of the piece.

Slow practice also uses the muscles in a completely different way, so we are not practicing what we want to do physically, either. This seems obvious from observing one's own body while playing, but it's nice to have scientific confirmation. Dr. George Moore, a researcher into the neurophysiology of musical performance, formerly of the University of Southern California, says slow practice "uses the muscles in a completely different pattern" from playing at tempo. In slow practice, Dr. Moore says, "You're not training your muscles to play the passage, period."

Slow practice tends to make us feel, think, and act in units of individual notes, which puts us at a further remove from the composer, who conceived the piece as a whole. One of many problems this creates is that, as we add up notes to make phrases, our timing errors also add up.

And when we have "mastered" the piece in slow practice comes the horrible moment when we say to ourselves, "Well, I can play it slowly. Now how do I get it up to tempo?"

Outlining takes the other possible approach. Too many things to think about, in too little time? Very well then; instead of *increasing* the amount of time, *decrease* the number of things to do; that is, play fewer notes.

But we do not omit the other elements. Tempo, dynamic shaping, phrasing—we play a complete performance. We can tape this performance and do everything else we do with a "noteful" version.

Why Outlining?

Do we outline just so we can tape? No, we do it because it lets us perform a piece with complete emotional involvement from the first moment of learning; *so we are always practicing what we want to do.*

Since we outline at performance tempo, and since tempo is crucial to the emotion, the emotion can be right. And since the dance of our playing begins at tempo, our bodies can learn the physical mechanisms necessary for the large-scale picture. *This* downbeat note is *here*, very loud. The downbeat two bars farther on is *there*, very soft. As we gradually "tuck in" the in-between notes, the physical means we use to play can develop "from the top down."

And since the dance is of notes rather far apart in time, if each note has the same timing error as in slow practice, it will yet affect the piece much less; for the error will be spread over the longer interval between notes of the outline. The errors will be divided down instead of added up.

And we never have that horrible moment when we ask, "Now how do I get it up to tempo?"

Survival of the Masochistic

We all play slowly sometimes, but why? Perhaps to hear more clearly what's happening melodically or harmonically in a particular passage; perhaps just to "feel our way" into it.

But we should not take slow playing as itself being practice because (as I hope you are convinced) it cannot be a true practicing of what we want to do.

Then how did it reach its current dominance, where it threatens to push out other forms of work? Perhaps the gifts of great players led lesser players to mistake the usefulness of slow work. (A gifted individual, inherently more efficient than the rest of us, can survive all kinds of inefficiencies in work methods.)

Errors will be divided down instead of added up.

As things are now, I sometimes think the real purpose of slow practice, and many other methods in the usual sort of teaching, is simply to screen out students who are not motivated. A cynic might say it's to screen out those who are not truly masochistic. Slow practice is an evolutionary pressure whose real function is hidden from everyone, even the teachers who promote it.

Session 11

Outlining

What notes should be in the first outline? And what notes should be added at each step so we finally play the piece as written?

A simple rule that works surprisingly often is to use the downbeat of each measure, or the notes on the beats. (See Illustration C in Session 9.)

The more general rule is that your ear decides what notes shall be present in any outline. Pick those which seem musically most important. I show below two outlines of the opening of Beethoven's (piano) Sonata, Opus 2, No. 2, as examples. One might be your first outline; the other, a substantially later outline of the same passage. (Until you are experienced, you may find it helpful to mark outline notes on your score with a highlighting marker.)

The first outline must have so few notes that playing does not interfere with the dance at all. This will be fewer notes than you think you could manage. Play the notes short, but not staccato; they are simply points on a curve. You are constructing a "follow-the-dots" version of the piece.

The final outline is the piece as written.

PLAY the outline as a performance, with complete commitment.

TAPE the performance and do all the things we discussed in previous sessions. Does it feel ridiculous with so few notes? I know how you feel; but you will be surprised at how the piece comes through even the simplest outline, and at how busy the performance keeps you.

Now ADD some well-chosen notes for the next outline. As you play, notice whether the added notes disturb your body's carrying of the phrase. If so, drop back a level. If you are uncomfortable even with the first outline, make a simpler one.

There is nothing monolithic about any outline. You may outline more notes per bar in easier places and fewer in more difficult ones. In a later outline, you may add more notes in some places while dropping notes in others. The rule is simply never to interfere with the continuity.

Continue in this way. The final outline is the piece as written.

Illustration D1. *Possible first outline for the opening of Beethoven's Piano Sonata, Op. 2, No. 2.*

Illustration D2. *Possible third or fourth outline of the same passage.*

"You Can't Practice It"

But don't some difficulties require separate work? Yes, of course; and if you are a pianist, again I urge you to read Whiteside for a wonderfully practical discussion of how one should actually play the piano.

For help on an even more fundamental level, learn the "Alexander Technique." This is not specific to playing musical instruments but is a general technique for using your body optimally in everything you do. Discussing it is beyond the scope of this little book, but I will say that its simplicity and power are amazing, and that at my first Alexander lesson, which took place at the piano, I played better than I ever had before. You can find a qualified Alexander teacher in your area by calling the North American Society for the Teachers of the Alexander Technique at (800) 473-0620, or Alexander Technique International at (202) 362-1649. In Appendix A you will find listed an excellent book on the subject by my friends Barbara and Bill Conable.

"Oh my dear, you can't practice it. Either you can play it or you can't."

Outlining and singing make many difficulties vanish so readily that it can be a little unnerving. And our problem with the ones that don't vanish is sometimes not the difficulties, but their significance for us.

One evening long ago, as I was coming up the steps of the Longy School on my way to practice, I ran into a young woman I knew, a student of accompanying. I saw she was almost in tears, and I asked what was wrong. She told me she had heard Schubert's famous song "The Erl-King" performed the previous afternoon at Jordan Hall in Boston, and that Madame Thus-And-So had played the notoriously difficult accompaniment fantastically well. My friend had gone backstage afterwards and approached the lady.

"Oh, Madame, your Erl-King was fabulous. Please tell me how you practice it."

"Oh, my dear, you can't *practice* it. Either you can play it or you can't."

What bothered my friend, I think (besides Madame's nastiness), was not her inability to play Erl-King. Why should it bother her to be in the same category as most pianists? What bothered her was that she took Madame's insensitive remark to mean, "You are not one of the elect," and *she accepted that judgment.*

Sometimes we do to ourselves the same thing Madame did to my friend, as when we know a way to overcome a technical difficulty but we regard that way as not "legitimate." We feel we ought to be able to play the passage in some other, more "legitimate," way. Thus we exclude ourselves from pleasure and achievement. To avoid this, commit the following to memory:

In solving musical problems, be idealistic.
In solving technical problems, be opportunistic.

Part IV

Recording Equipment

We can dance, we can sing, we can spot-check with any equipment, no matter how bad its sound; so begin taping right away, even if your equipment is poor and you cannot afford to replace it. Session 12 tells you how to set up and use your equipment.

Session 13 explains the role of each component and recommends specific models in case you do want to buy new equipment.

Sessions 14 and 15 are about tapes by which others will judge our playing: audition tapes and commercial releases. When we listen to our own tapes, we hear our intentions and ideals; when others listen, they hear only our sound. We want them to hear it as accurately as possible, because a bad recording affects the listener's judgment of the interpretation. Session 14 is a "pep talk" about this, while Session 15 gives specific advice.

Recording equipment is no harder to understand than the musical equipment we use every day, that is, our instruments. There is no need for anxiety. Treat this like a smorgasbord, and take what pleases you.

Session 12

Low-Anxiety Recording

In which I suggest how to set up and use your equipment, assuming you already have equipment and generally know how to operate it. The instructions apply to both conventional (analog) and digital recorders.

If you do not have equipment, the next session recommends some. If you do not know how to operate it, ask a knowledgeable friend or take the final desperate plunge and read the manuals!

1. Place the tape deck so that, from your playing position, you can see its meters and reach its controls easily.
2. Put the microphone three or four feet from your instrument and aimed at it.

   Later, you may wish to try the microphone at various distances, aimed at various parts of the instrument.
3. Turn the volume control all the way down to avoid "feedback."

   Volume control means the control for the loudspeaker volume. When I mean the control for how strongly the sound is recorded on the tape, I will say "record level" (re-CORD level) control.
4. Plug the microphone cable into the tape deck and switch the deck to "microphone input" (rather than "line input").

   If your deck does not have a microphone input, you will need another unit called a "microphone preamp." In that case, plug the microphone into the preamp and the preamp into the deck's line input. (See Session 13 for more about microphone preamps.)

   I will assume you are using just one microphone, which is all you need for anything in this book. Not only is one less expensive than two, but using two microphones incorrectly can blur your attacks and alter your sound. (For the correct use of two microphones, see Session 15.)
5. Put a blank tape into the deck.

   Always use a reputable brand: TDK, Maxell, and so on; not an unlabeled or retail-store "house brand." Replace the tape with a new one every so often.
6. Reset the tape counter to 0.
7. If your recorder has Dolby B noise reduction, turn it on.
8. Start the tape deck recording. ("Put the deck into record mode," as they say in the manuals.) Play or talk and adjust the tape deck's record level control so the meter occasionally goes above 0.

   If a cassette deck refuses to record, probably a tab has been removed from the edge of the cassette shell to protect a recording already on the tape. To allow recording, put a small piece of adhesive tape over the hole.

If a digital audio tape (DAT) won't record, you will find on the edge of its shell a slider which you can slide to its other position.

The DAT has just the one slider. The cassette has two protection tabs, one for each direction of tape travel, or each "side" of the cassette. To record on one side, cover the hole on the end from which the tape feeds while recording that side.

Don't erase a recording you need!

9. Rewind the tape until the counter reads 0.

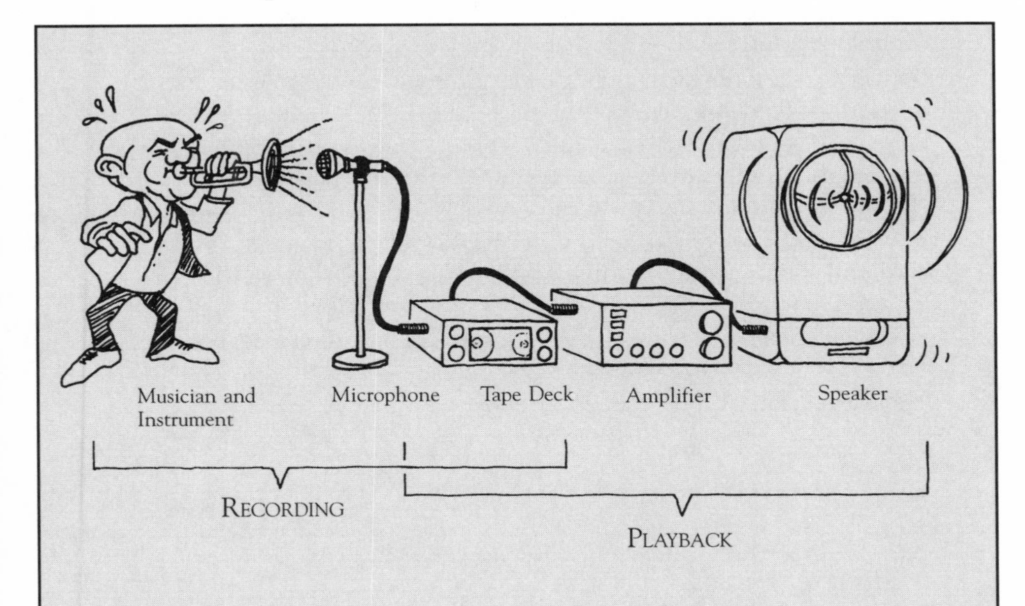

The instrument makes sound waves. The microphone turns the sound waves into electrical waves. The tape deck stores these and plays them back on demand. The amplifier makes them powerful enough to operate the loudspeaker. The loudspeaker turns the electrical waves back into sound waves. (Recording and playback are shown in the same illustration for convenience. Do not record and play back at the same time!)

Illustration E. *A Musician Playing and Recording*

10. If your deck has a switch labeled "Source/Tape" or "Input/Tape," put it into the "Tape" position.
11. Put the deck in Play.
12. Turn up the volume control.
    This may be a control labeled "playback level" or "output level" on the tape deck, a volume control on your amplifier or receiver, or both.
13. You should now hear the sound you recorded.

14. When recording, adjust the Record Level control so the meters go up to about +3 in the loudest passages.

This is just to get you started. As you continue to work, if a recording sounds distorted, play it again at lower volume. If it then sounds undistorted, the distortion was not on the tape, but was caused by the amplifier or speakers. The solution is simply to listen at lower volume (or to get different speakers, a different amplifier or both).

If the distortion is still there when you play the recording softer, then it really is on the tape. The solution is to reduce the record level and make another tape. If the recording now sounds clean, leave the level control there for a while.

Conversely, if the tape never sounds distorted even on your loudest passages, you can raise the record level.

You will gradually become familiar with the correct setting, a compromise between overloading the tape with your fortissimos, on the one hand, and losing your pianissimos in tape hiss, on the other hand.

While an incorrect setting may give bad sound, be reassured that it is very unlikely to damage anything. (But playing the tape louder than your amp or speakers are capable of can damage the speakers.)

Do not change the record level setting during a performance, because that will falsify your dynamics. Find the right setting for the very loudest passage, and leave it there for everything.

Session 13

Audio Systems and Components

We can do much with poor equipment, but ultimately we recognize that the beauty of the sound is not a "feature" riding on top of the music, but an integral part of it, the sensuous bridge to the meaning. When it comes to the subtleties of performance, we do want to hear just what we are creating.

Accurate-sounding equipment is costly. On the other hand, not all costly equipment is accurate. Sound quality varies widely at any given price, whether that price be low or high. Thus, it is worthwhile to choose carefully.

But what equipment should you try, and how much will it cost? To answer this, I put on my audio consultant's hat and carefully evaluated a variety of units. Many were worthless, and I don't even mention them here. The ones I do list are worth your attention; those in ***boldface*** *offer unusually good performance and/or value for money.*

Certain products have long lives (the Coles microphone has been made since the 1950's, the LS 3/5A speaker since the '70's), but most don't; and prices change almost overnight. Nevertheless, this snapshot of what was available in late 1995, and how much it cost, should be useful as a point of departure.

The beauty of the sound is the sensuous bridge to the meaning.

THE RECORDING SYSTEM

MICROPHONES: PARADISE LOST CANNOT BE REGAINED

The microphone turns your sound waves into electrical waves. In doing so, it must deal simultaneously with a wide range of dynamics and a wide range of pitch. It must handle your fortissimos with aplomb, yet not ride roughshod over your pianissimos. It must be "tonally neutral," treating your entire pitch range and the highest overtones evenhandedly, not favoring any particular note or register.

All audio equipment has this double task, but microphones are extra-important because they come first in the recording chain. The beauty they corrupt can never be made intact.

Good microphones are expensive, and if you cannot afford any of the ones listed below, get whatever you can afford. Remember that you need only one.

I have heard the following models:

Audio-Technica 813A ($208)
Shure 809 ($228)
Beyer M260 ($399)
AKG C-1000S ($429)
Shure SM90A ($310 each + $165 for power supply adequate for two microphones)
AKG SE300B/CK94 ($618 each + $110 for power supply adequate for two microphones)
Coles 4038 ($1045)

Short on Cash?

Components recommended here include some good inexpensive models but only a few really cheap ones. Don't let this stop you if your budget is small.

You may already own a stereo system with a cassette deck. Then just buy one microphone (and a mike preamp if your deck needs it), and you're ready to go.

If all else fails, you can buy whatever is on sale at your local stereo or musical instrument store. But there are many alternatives. Borrow equipment, or find excellent used equipment for low prices in your classified ads or at garage sales. (See the discussion of the Dynaco Stereo 70 amplifier. I would be wary of buying mechanical-type equipment like recorders and turntables used, however.) Cycling to school one day, a student of mine even found a superb receiver in perfect condition thrown out with someone's trash. Cost: $0.00.

Whatever your outlay, put it in perspective. Compare it to the cost of a year's lessons if you are a student, or the income from a year's lessons if you are a teacher. If you use the equipment in the ways we have discussed, it will hugely enhance your work. So please—start today!

The Shure 809 and SM90A sit unobtrusively on the floor, saving the cost of a microphone stand. Of these two, the SM90A is tonally more neutral and more "open"-sounding and costs twice as much, but the 809 is very usable and convenient.

The Audio-Technica sounds musical and has plenty of bass, but sounded slightly "woolly" to me, with some overtones slightly exaggerated.

The AKG C-1000S is pleasing and clear, although it too exaggerates overtones somewhat and is light on bass to boot. If this is bothersome—which depends on your instrument, room, speakers, and personal preference—try bringing the mike closer to the instrument. (If you do this with the Beyer or the Audio-Technica, you will have too much bass.)

The AKG SE300B/CK94 is a professional-sounding mike with fine clarity, good bass, and good all around performance.

The Beyer M260 is a standout, with a dynamic sound that is tonally quite neutral.

> ***Let Our Voices Be Heard***
>
> Why isn't there more cheap equipment suited to our needs? There's no technical reason we couldn't have a selection of $100 microphones, $350 recorders that would work directly from those microphones, $300 amplifiers and $400 speakers which would recreate our sound pretty well. Yet right now our choices are few or none.
>
> The reason is that the big manufacturers are oriented toward sales in the hundreds of thousands, and the great mass of their customers are not tuned in to accurate sound. Some microphone and speaker makers told me that their products have intentionally inaccurate sound to come across better in showrooms.
>
> Let's change this situation! Drop me a note saying exactly what *you* would like to buy that is not available. Address me by E-Mail at boyk@caltech.edu, or in care of Performance Recordings, 2135 Holmby Avenue, Los Angeles, CA 90025-5915. I will use your note to help convince companies to make the equipment we need.

The Coles, designed by the British Broadcasting Corporation (BBC), is the best microphone I have ever heard.

There are a zillion microphones in this world. Most aren't very good. Microphones whose specifications look promising, but which I have not heard, include, in order of increasing price: Shure 849 ($211); Shure SM94 ($280); Shure SM81 ($607 including power supply); and the Sennheiser 441 ($895).

The store selling you the microphone can also sell you a mike stand and a shock mount. Except for the Shure 809 and SM90A, you need both.

To find the nearest dealers for these and all other products listed, see Appendix D for distributors' phone numbers.

Recorders

The ideal recorder is inexpensive, works straight from your microphone, and sounds decent. Unfortunately, I can't find any like this.

The portable ***Sony TC-D5M*** ($799.95, but $599 from CAM Audio, see Appendix D) works beautifully straight from any microphone, sounds excellent, and meets the need perfectly except for being pretty expensive. The price is better than it looks, however, because this machine contains a small amplifier and speaker as well as a headphone jack.

The discontinued ***Technics RS-BR 465*** ($240, street price $190) works well with most microphones. Its "wow and flutter" (unsteadiness of speed) adds a little roughness to the sound, but not too much for practice purposes. You might still find one of these units in a store.

Tin Cans and Tape Heads

I said earlier—in the illustration at the beginning of Session 12—that the tape deck stores electrical waves and plays them back on demand. Actually we have no way of storing waves of electricity, but we *can* store waves of magnetism, electricity's alter ego. We do this by "printing" the moment-to-moment changes of magnetism onto millions of magnets glued onto a moving tape—so many magnets, and each so tiny, that the tape surface looks continuous and smooth. The device which does this printing, and which also converts the electricity into magnetism, is called a *recording head.*

Just as the tin can at either end of a child's tin-can telephone serves in turn as sender and receiver, so a recording head may also serve to play back the tape; but for best performance, two separate heads are used, the *play head* following the record head. This allows one to hear the recording off the tape as it is being made, which is not possible with a combined *record/play head,* any more than one can talk and listen at the same time on a tin-can telephone.

Whenever a deck records, an *erase head* first scrambles any previous recording.

In turn, then, any point on the tape passes over the Erase, Record, and Play heads in a "three-head" recorder; if Record and Play heads are combined in a single head, we have a "two-head" machine.

Other inexpensive units I've tried either have too much wow and flutter or do not work well from microphones. (Makers sometimes provide places to plug in microphones on machines which in reality will not record well from microphones, but only from an electrically stronger source like an FM tuner or CD player.)

Microphone Preamplifiers

You can use any recorder you choose if you buy a "microphone preamplifier," a separate piece of equipment which boosts the microphone output. A microphone preamp allows you to use the recorder you already own, or to choose your new one from the full range of those available.

The ***Symetrix SX202*** ($299) is a two-channel microphone preamplifier, so you can use it for stereo recording.

The *RTS 1400* ($285) is a one-channel unit, and battery-powered. For our work, you need only one channel, but if you buy a single unit now, you can get an identical one later for stereo recording.

I have not heard these preamps, but from their published technical data, they will make tapes as quiet as your recorder is capable of, no matter what microphone you use with them.

> ***Microphone Preamplifiers***
>
> If you plugged a microphone into a tape deck not intended for recording from microphones, you would find yourself taping mostly the noise of the deck's own circuitry. It would be like listening to a weak radio station and getting mostly static: You can play the radio as loud as you want, but making the station louder makes the static louder, too. The difference is that in recording, the "static" comes from within the tape deck itself.
>
> To overcome it, you need a piece of gear to strengthen the electricity from the microphone—*amplify* it, we say—*before* it gets to the tape deck, making it strong enough to drown out the deck's own noise. We call this unit a *preamplifier*, a "before-amplifier"; and of course the "preamp" itself must not add more noise than it can help. (Adding some is unavoidable in this Universe.) Since preamplifiers may be used in other parts of an audio system, we specify that this one is a *microphone* preamplifier. Thus we want a *low-noise microphone preamplifier*.

Digital Audio Tape (DAT) Decks

The tiny portable *Sony TCD-D8* ($899 list, but $677 from Broadcast Supply West, see Appendix D) includes microphone inputs. I don't know how usable they are, nor how good is the sound quality of the machine itself.

THE MONITOR SYSTEM

So far you have picked up your sound with a microphone and transferred it to tape. What remains is to play it back with an amplifier and loudspeaker, the "monitor system," as we say.

Perhaps you already own this equipment, and you can play back your tapes the way you would play prerecorded ones. Don't be surprised, however, if you find that your tapes demand more amplifier power, and more loudspeaker ability to use that power, than commercial recordings.

Even in a modest room, a hundred watts or more of power may not be out of place, not because you want the music so very loud, but because each musical attack can require 25, 50, or even 100 times more power than the average. And your tapes may well capture those attacks better than commercial recordings.

Since our work needs only one microphone, you need only one loudspeaker and one amplifier. However, almost all amplifiers and receivers are stereo units. Speakers, too, come in pairs (as friends of ours found out when burglars stole *one* of theirs).

Loudspeakers

I have found two inexpensive models that are outstanding value for money, the tiny ***KEF Coda 7*** (list price $220/pair) and the small ***NHT 1.1*** ($380/pair). Both have a lively, "open" sound.

The Coda 7 has a slightly "hollow" quality on certain instruments, but I find it far more listenable than the same maker's Q-10, which sells for $350.

The NHT sounds more neutral and "open," and has more bass, though it is still not very powerful down there. (No "sock in the gut"; to get that, multiply these prices by six or so.)

The NHT does better than I had thought possible at the price. It even gives the legendary—and much more expensive—LS3/5A (see below) a run for its money. (The Caltech Music Lab, which I direct, owns a pair of the NHT speakers bought at a special "academic" price, but this has not affected my judgment of their quality.)

The bookshelf-sized B&W 610i ($500/pair) offers more bass and power than the NHT, and is quite neutral tonally. A usable speaker, but I find it lacking in dynamic life and "openness."

For this book, I listened to four other models at or below the price of the 610i, but found none of them recommendable.

If you do not buy one of the above speakers, then as one musician to another, I urge you to hear a tiny model called the LS3/5A, designed by the BBC and manufactured under license by three companies. This speaker will not give you as much bass as the B&W 610i, nor can it drive you out of the room with high volume levels, but its musical accuracy is remarkable. It is expensive, but provides a useful reference point for evaluating other models.

At this writing, the ***Harbeth LS3/5A*** ($1100/pair), the ***Spendor LS3/5A*** ($1250/pair), and the ***KEF LS3/5A*** ($1450/pair) are available in the USA. There's no reason to buy any but the least expensive of these.

The ***Spendor SP-1*** was an outstanding speaker with the natural sound of the LS3/5A but with more bass and higher maximum volume. On my recommendation, the famous audiophile label Sheffield Lab took a pair to Russia as on-location monitors, and were pleased with the results. Now discontinued, the SP-1 has been replaced by the SP-1/2 ($2400/pair), which despite the similar name is quite a different design. I have not heard it.

Unless the instructions specifically say otherwise, any speakers will benefit from being put up on stands and well clear of walls.

Amplifiers and Receivers

These categories change too fast to keep track of, so I will mention some I have heard and some I've only heard about. Two useful rules are:

> *Choose an amplifier whose power is at or near the upper limit of the recommended range for your loudspeaker.*
>
> *Try the amplifier with your speakers. As with musicians, some ensembles work well and others not.*

An outstanding "integrated amplifier" (Illustration F) was the **NAD 3020** (20 watts per channel; discontinued; was $180). You might find a used one in good condition.

A usable current model is the *Rotel RA970BX* (60 watts per channel; $479). The sound is fair for the money, and much superior to one competing model I auditioned. Compared to the NAD, however, the Rotel, while somewhat more powerful, is not to my ears so "open," dynamic and listenable. In addition, the Rotel has no provision for playing Lp's, while the NAD plays them very well.

The *Hafler XL-280* (140 watts per channel; discontinued; was $799) was a fine amplifier whose sound never offended, though it seemed to me somewhat lacking in dynamics and presence.

Some amplifiers have "input level" controls that can be used as volume controls. A few of these amps are sensitive enough to give full power when connected directly to your recorder so you don't need a preamp (unless you want to play Lp's).

The *Adcom GFA 555 II* (200 watts per channel; discontinued; was $950) was fair value for money, powerful and dynamic, with good bass. It has been dropped in favor of a new model which I expect to sound completely different.

The amp which has brought more high-quality sound to more people for the least money is the ***Dynaco Stereo 70*** (35 watts per channel), a tube amplifier still available used. A friend bought one in perfect condition for $65—including its matching **PAS-3** preamp!

If all this is too confusing, just listen to some of the recommended units and choose the ones that please you most.

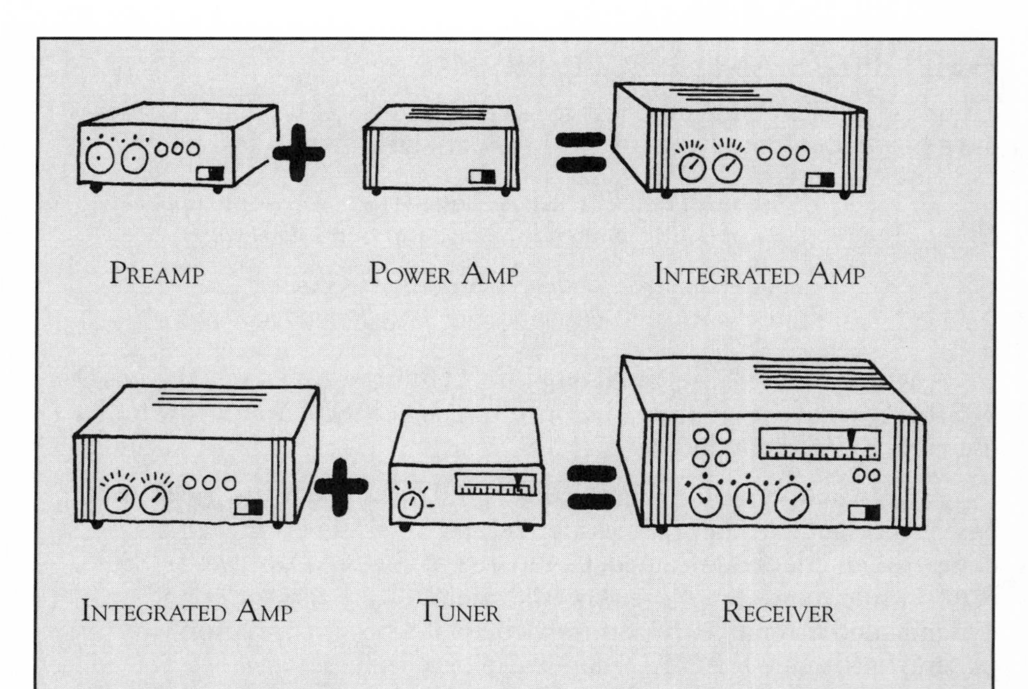

The *preamplifier* is the part of your stereo system with the volume control; a switch to select FM, CD, etc.; perhaps bass and treble controls; and so on. It may be separate from or on the same chassis with the *power amplifier* and/or the *tuner*. (Don't confuse this preamplifier used for playback with the microphone preamplifier used for recording.) The power amplifier provides "muscle" to move the loudspeaker. Often it has no controls except an on-off switch, though some power amps do have an "input level" control, another name for a volume control. (Sometimes this lets you do without a preamplifier.) An *integrated amplifier* is one chassis combining preamp and power amp. A *receiver* is one chassis combining an integrated amp and a tuner.

Illustration F. *Relationships of Components.*

Session 14

Musicians and Magic; Or, Why Pitchers Can't Hit

In which we go on a psychological excursion but work our way back to audio.

Some pianists deny the power of the piano itself to affect the art they create. They say, not just that a bad piano doesn't bother them, but that they believe it does not affect the music reaching the audience.

It's true that some important aspects of the music will survive a bad instrument. TIME elements, for instance. And ARTICULATION may survive in some measure, depending on just what the instrument's flaws are. But DYNAMICS, SONORITY, BEAUTY—these and other aspects may well vanish as flexible contributors to an interpretation, and with them much of the music.

The air was clean, the food was wonderful, the quiet was perfect.

So why do these pianists deny the vulnerability of their art? They turn to this magical thinking out of impotence about having to play on bad instruments.

I remember thinking this way myself when I played on a clunker of a piano at a college in California. This college was an unusual place. It owned the beautiful mountain valley in which it was located, and ran the valley as a farm and ranch staffed by the students. The food was just-picked, plain-cooked and wonderful. The air was so clean you could smell a single car crossing the valley on the through-road miles away. The quiet was perfect. There was simply no background noise at all; and my wife and I could feel our hearing getting more and more sensitive the whole time we were there.

When we first arrived at the tiny group of college buildings, I started walking toward the one which I had been told housed the piano. A student ran up to me. "Are you Mr. Boyk?" he asked. When I said yes, he continued,

"I'm supposed to tell you not to go in there yet."

"Why not?" I asked.

"Because Larry's gone to town to get some dowels for the pedals," he replied. "They won't work 'til he gets back." (Larry was also a student. The dowels were to replace missing metal rods which normally convey the pedals' action up into the body of the piano.)

I assured the student I would not faint at the sight of non-working pedals, and as we went into the building, he informed me that Larry would also tune the piano when he returned. "We had a tuner here last year, and Larry watched him."

The admirable Larry's tuning was the best thing about the piano. Many of the hammers were bald, the felt worn down so far on top that what hit the strings was the wooden core of the hammer. These notes sounded like a barroom piano, but at least they had character. The rest of the keyboard sounded pallid. And the soft pedal, I was unnerved to discover during the concert, shifted the keyboard to the *left*, unlike any other I've used.

"My commitment and art will come through," I said to myself as I performed. "My focus will triumph." It was baloney and I knew it; but it got me through.

As she sat down at the piano, she realized it was a spinet with a grand piano-shaped table behind it!

I wonder if the same self-delusion worked for a pianist I know of, who arrived at the last second for her noon concert at a ladies' club in the South. Walking onstage, she wondered why the piano lid was down and closed, with a cloth draped down to the floor all around.

As she reached the instrument and sat down, she realized the awful truth. The piano was a *spinet* with a grand piano-shaped table behind it!

The fantasies that say we can overcome such situations are helpful in maintaining our sanity; but when they lead us to be slack about assuring top condition for the instruments we perform on, then we have too much invested in magic and too little in reality.

Another group with too much investment of belief in the wrong place, I have always thought, is baseball pitchers, whose weakness as batters is legendary. Their faith in pitchers' omnipotence extends even to opposing pitchers; hence their own inability to hit.

We pianists are unfortunate in not carrying our instruments around with us. But we are not the only ones with omnipotence fantasies; for many musicians feel their art will come through despite defective recording or inadequate playback.

This is wrong, as I found out when I sent my first album, an Lp, to a young pianist friend who was studying at Peabody Conservatory. In her return letter, she was enthusiastic about the Scarlatti; but her embarrassment was almost palpable when she discussed the Beethoven Sonata Opus 111. She said the first movement seemed to have too many climaxes, and too much banging.

I did not hear the banging or the multiple climaxes. But I knew I might be hearing what I thought I had done, or what I had wanted to do, not what I really did. I decided to listen again in a few months.

The summer came before I got around to it, and my friend wrote again. At home in Colorado for the vacation, she had listened on her father's system, which was much better than her dormitory-room player.

Now she did not hear the extra climaxes or the banging. Now she loved the performance. But of course she was puzzled by the change from the same disc heard at school.

Reading her letter, I suddenly realized what had been wrong. Unlike most recordings, this one had a wide dynamic range. When the music got above, say, *forte*, it had overloaded her dormitory system. Any such passage came out equal in loudness to any other such passage. Hence the multiple climaxes. They were graded dynamically in the playing and on the recording, but couldn't be distinguished by the system.

She heard the ugliness created by her stereo system as though it were created at the piano.

Overloading also makes the reproduced tone ugly; but because my friend was thinking in musical terms, she heard the ugliness created by her system *as though it were created at the piano*.

This shows why record reviewers who don't have good systems risk misjudging interpretation along with sound quality. Unfortunately, reviewers are mostly impoverished journalists ignorant about audio; so their systems are poor.

Most review magazines do not want to address this issue. The editor of an influential one told me, "I don't want to open that can of worms," after his magazine ran two reviews of one of my albums, by two different reviewers, in the same issue. One said the sound was superb, the other called it bad. The editor refused to tell what equipment was used by the reviewer who panned the sound. "A matter of principle," he called it.

Session 15

Counsels of Imperfection

Adapted from a letter I wrote to a young professional musician. He had asked for advice on recording for commercial release.

"In Some Darkened Room"

Dear David,

Thanks for your letter, and congratulations on your recent performance prize. You ask about "producing a recording in the modern world." Merely by asking, you put yourself in the select group of performers who take responsibility for seeing that the music reaches the listener's ears in good condition, whether the listener is in the hall with you or at home listening to your recording.

I know the feeling you mean, that after we are done "making our noises in some darkened room," as you put it, "the evidence vanishes." I think the fundamental problem is not that our music-making is soon gone, it's that *we* are. "Making our noises in some darkened room" is a metaphor for life. The darkened room is the Cosmos.

You comment that visual artists do not have the problem of preserving their work. But they do! Some of Van Gogh's colors aren't anything like the originals, because he used unstable pigments. And remember the Sistine Chapel restoration, and the fuss about how those frescos looked?

For us musicians, preserving our art is only one use of recording. I'll come back later to some others.

The easy way to make a recording is to hire a producer to do it all for you, so you only have to play—and pay. The producer would see to recording and editing the tapes; transferring them to cassette or CD; writing program notes; overseeing design, typography and printing; and so on.

If you ever find a person for hire who does a good job of all this, let me know! I don't think it can be done, because customers—that's you—will not be willing to pay the cost of a good job plus profit to the producer.

Instead, you will have to deal with the recording engineer yourself. You will find that some engineers are also musicians, or at least have a real love of music. Others don't know or care anything about it. Some have the same drive to perfection we performers have; others are careless. But no matter who your engineer is, *you* must know something about the matter, because yours is the judgment which must prevail. You are the musician. It is your playing, your art, your instrument. You must decide what is good and what is not.

The Ear of the Beholder

The crucial knowledge is ear knowledge. Your ears will tell you what works and what doesn't. Just listen and judge the sound *as though it were live music in front of you*.

You will find it never passes this test. Never hire a recording engineer who says it will, because he is not just ignorant but dangerous. The best recording is much closer to live than the worst, and worth pursuing; but it never sounds live. That is why my recommendations are "counsels of *im*perfection."

Judge the recorded sound as though it were live music in front of you.

Though ear knowledge is what you need, it will help to have some familiarity with the jungle of recording equipment and techniques. In what follows, I point out trails that lead to the goal, not to turn you into a recording engineer, but so that you can hire an engineer who shows some familiarity with the correct paths.

Opinionated Answers

A short answer to your question is that it's easy to make a good recording. Set up a pair of Coles 4038 ribbon microphones in a particular arrangement I'll describe later, in an acoustically good hall where you are giving a concert. Plug them into the microphone inputs of a Revox B-77 tape recorder. Use Ampex "Grandmaster" (Type 456) tape. There you are!

A longer answer is that the following are important:

(a) Record in concert, or at least with audience.
(b) Use a microphone arrangement which gives convincing stereo.
(c) Use ribbon-type microphones.
(d) Record on two-track analog tape.
(e) Trust your own ears.

Let's take them in turn.

(a) WHY RECORD IN CONCERT? Because you are trying to record your art, and that art does not consist simply of playing the clarinet, but of using the instrument and the music to communicate emotion. Communication requires two parties, you and the audience; so record with an audience. And since communication is heightened by a concert's sense of occasion, record in concert.

Personally, I tape two concerts in the same hall. If I can't get a releasable recording with just a few edits between the two, I conclude that I wasn't ready to release the piece anyway.

"My artists are not comfortable recording in concert," a producer for a big record company said to me. I wanted to ask her what *comfort* has to do with it! When we give our most moving performances, are we comfortable? When we hear a performance so great it makes us ache, do we describe it as comfortable? Her remark was symptomatic, I think, of something fundamentally wrong with the way most recordings are made.

If you don't record in concert, you should still record with an audience and in long "takes." I did the Prokofiev Sixth Sonata this way, because the concert tape was spoiled by a technical problem. "Take one" was the entire work; take two, the first movement again. Then we did three or four takes as "covers" for a few flubs. Others were left as is, because editing always carries a risk of disturbing the musical line.

Most people have never heard true stereo.

"Should I record this concert at all?" you may ask yourself. "By recording, am I saying I think this will be my best playing, worthy of preserving?" Questions like these sap energy, so a long time ago I decided always to record.

(b) IN TRUE STEREOPHONIC SOUND, the musicians appear to have well-defined locations both left-to-right and front-to-back; and they remain in these locations despite changes in pitch or dynamics. As in live sound, you can turn your head toward each instrument.

You will realize from this description that most recordings don't have any such convincing "image" at all. And you are right: Most people have never heard true stereo.

Does this matter? Yes, because locating sounds is fundamental to us as human beings, with an importance rooted in our evolution. Whether we were the predators or the prey, we needed to know where other animals were; and this is still a part of us. Notice, for example, how you automatically face any sound you are attending to, and instinctively cock your head to locate an unexpected sound.

I think we must satisfy the "evolutionary being" in us with good imaging before we can address our "esthetic being" with good sound quality.

Luckily, good imaging is easy with the proper choice and arrangement of microphones. (See Illustration G.) Some of my students and I made an easy-to-understand album of this, available as the Demonstration of Stereo Microphone Technique. (See Appendices A and B.)

As you will hear on that album, a very good image may be created this way: Take a pair of microphones which are sensitive to front and rear but not to the sides ("bi-directional" or "figure-8" microphones). Put them head to head, one exactly above the other, but aimed at right angles, bracketing the stage. (This is called a "Blumlein pair," after Alan Dower Blumlein, the man who invented stereo.)

Directional Sensitivity of "Figure-8" Microphone—Top View

• Size of pianos indicates sensitivity; Full to front and rear, falling to half as sensitive in directions marked; and insensitive to the sides.

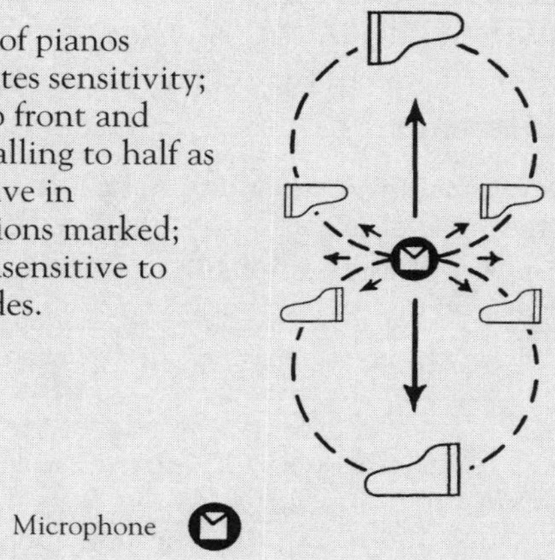

• Such diagrams as these show sensitivity vs. direction. The *size* of the diagram means nothing; the *shape* everything.

Microphone

Blumlein Pair As Seen From Center Stage

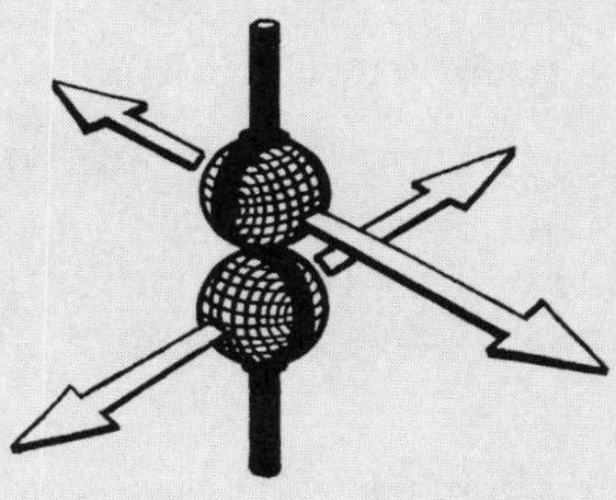

Top View of Blumlein Pair Patterns

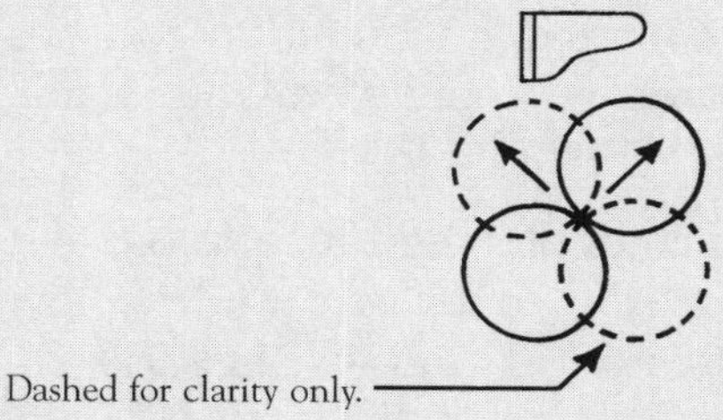

May be used on any ensemble.

Dashed for clarity only.

Illustration G. *Figure-8 Microphones and Blumlein Miking.*

(c) USE "RIBBON" MICROPHONES. Professionals in recording commonly use "condenser" microphones. Some expensive condensers sound good, and some ribbon units sound bad; but the best ribbons give the most musical sound I have heard.

The best microphone I have ever heard is the **Coles 4038**, from England, made to BBC specifications. The long-discontinued ***B&O 200*** (which sold originally for $129.95!) is also superb, and convenient for stereo because it has two complete mikes in one physical assembly. (Don't buy a used B&O without being positive that it is in perfect condition, for damage to the ribbon cannot be repaired.)

> In stereo miking, a Blumlein pair is optimal only if you have a pair of bi-directional ribbon microphones. If you have ribbons which are not bi-directional, it can't be Blumlein, by definition. If the mikes are bi-directional but not ribbons, you probably shouldn't do Blumlein, for reasons beyond our scope here.
>
> In either case, your engineer may use one of several setups described by phrases such as "M-S pair," "ORTF," "coincident cardioids" or the like. Don't let the jargon throw you. Listen to the demo CD mentioned above, and read the booklet that comes with it. You will then know more about the subject than most professionals, and be able to sling around enough jargon yourself to keep at bay a whole control room full of engineers.

If you would like to hear how these things sound on actual music, all of my own records use a Blumlein pair on piano; and all except one use ribbons. A Blumlein pair of ribbons was also used for three Sheffield Lab albums with which I was involved. (See Appendix B.)

(d) WHY ANALOG? Because it sounds better than digital. And because you can convert your analog master to any existing or future digital format. If the master is digital, on the other hand, converting to another digital format can degrade the sound. Since formats are still in flux, analog masters are valuable.

Despite the virtues of digital technology, it is inferior to analog in reproducing the subtleties of music.

Many people run a backup recorder also, so they will still have a tape if the main machine fails. Many use DAT (digital audio tape) for this purpose.

You have heard of the supposed superiority of digital audio over analog. How does it happen that my recommendation goes against this, especially when my name is in the credits of several digital albums that have received rave reviews? (See Appendix B.)

Despite the acknowledged virtues of digital technology—stable pitch, easy duplication, convenience of use—to my ears it falls below analog in reproducing the subtleties of live music. I find textures and timbres homogenized in digital recordings, with dynamic inflection and ambience reduced compared to the original live sound or the control-room sound straight from the microphones.

This view is the result of careful comparisons of digital and analog sound in the Caltech Music Lab (which I direct) and in numerous recording sessions in which I have been performer, engineer, producer, or simply an interested guest.

Actually, what is unusual is not my view of the matter but simply that I am willing to speak out about it. Most of the people I know who are involved in the highest level of professional recording agree on analog's superiority but do not feel free to speak out.

It is interesting to compare the sound of a Compact Disc made from a digital master tape to one made from analog. Of course the final sound on any CD is digital, but it makes a difference whether that sound originated as digital or analog. On my CD of Mussorgsky's "Pictures at an Exhibition," you can hear the comparison for yourself. (And the Lp lets you hear a version that's analog all the way through. Its superiority in some important aspects of the sound is dismaying when you realize that Lp's are commercially dead.)

This recording was released in this expensive triple format precisely so that you could listen and decide the issue for yourself. At this writing there is no other recording I know of that lets you make the comparison.

The recording engineer's work may all sound bad to you. You will be right; it often is.

WHY TWO-TRACK? Because you are using only two microphones, and each needs just one track, whether analog or digital. The "multi-tracking" used for most commercial recordings degrades imaging and sound quality, while adding noise.

Many people don't know that two-track analog tape can be virtually silent. Listen some time to a professional recorder like an *Ampex ATR-102* or a *Studer A-80* running half-inch tape at 30 inches per second. You can turn the music up very loud and still hear just the tiniest hiss.

Even if you do have more hiss, such as the Revox might give you, it's not the end of the world. If it's constant, listeners "hear through" it. You can reduce or eliminate it with noise reduction methods, but these mean more electronic processing, which risks degrading the sound.

If you must use noise reduction, use a Dolby® system. But truly, there is no need for noise reduction in two-track mastering. I have never used it, and no reviewer has ever complained about hiss on my albums.

(e) TRUST YOUR OWN EARS! It's natural to be intimidated by the engineer's apparently expert knowledge. Do not give way to this! Judge the engineer's work by how it sounds to you. It may all sound bad. You will be right; it often is.

Judge the recorded sound as though it were live sound in front of you. If it sounds bad, say so. If you can't tell where the instruments are, if they don't sound like themselves, if the presentation changes when the sound gets loud or soft—or high or low—something's wrong. *If it sounds like a good recording, something's wrong. You want it to sound like live music.*

If the engineer can't do better, get another engineer. Do not waste energy trying to get him to work differently from the way he usually does. People work at a certain quality level, and it's useless to try to explain a higher level.

Ask questions in advance. If he starts telling you about recording a clarinet and piano recital with one mike on the clarinet, two on the piano and four more in the hall for ambience, do not hire him. If he says he will use a single "coincident pair" such as we discussed above, consider using him.

Cost

One question you will certainly ask is the cost. How much have you spent so far? Let's say you were doing the concerts anyway, and the room and other musicians cost you nothing. Then your cost so far is for the engineer and his equipment, and the actual reels of tape.

Say you do two performances of a three-reel concert. (A reel is 30 minutes usually.) The six reels of tape might cost $150. The engineer might cost anywhere from $400 to $1200 for two sessions. Adding these, let's say $1,000 total for the unedited master tape.

Could it be less? Of course. A single Digital Audio Tape (DAT) records for two hours and costs $10 or less. And you might find someone who records for a hobby and won't charge for his time. Or someone who's good but charges little. It might be $300. It might be zero.

> When recording for release, have the piano tuner on "standby" to touch up the tuning at intermission. Of course the piano must be tuned on the day of the concert, and touched up just before the house opens.
>
> Make sure the hall is as quiet as possible, acoustically and electrically. Turn off fans and air-conditioners, nearby shredders and copying machines, and so on. Most halls use a type of light dimmer called an SCR. These dimmers are like little radio stations broadcasting static. Make sure the engineer keeps the equipment physically away from the SCR's, avoids plugging into wall plugs on the same circuits they use, and uses filters for his power cords.
>
> Put a note like this in the program, too:
>
> ***We respectfully request*** *that you turn off all noisemaking devices, including alarm watches, telephones and beepers; and that you refrain from any noisy activity, including program rattling, candy unwrapping, and notewriting.* ***Thank you very much.***

Edit your master tape so it's exactly what you want your listeners to hear: not just the music, but also the spaces between pieces, the timing of the "room sound" coming up and fading down, and so on. (Room sound is the sound in the room with the audience present but silent, and with no music playing. Always record several minutes of it to use in your final tape. In a live concert, you may not be able to do this; but don't forget it in a session dedicated to recording.)

The Lp, with its well-known defects and little-known virtues, is all but dead.

Editing can cost anything from zero (if your hobbyist friend is good enough at it) to $1,000 if you need, say, 20 hours at $50 per hour. Perhaps five hours at $50 is a good estimate.

Compact Disc Cost Summary

This very approximate guide assumes that you record from concert tapes (see text.) One column gives a minimum cost (assuming you don't actually get it free); the other, a figure resulting from higher standards, bad planning, or both. It is not labeled Maximum because there is no maximum. Note how the cost per disc goes down for 2,000 discs.

Costs Independent of How Many CD's You Make

	Minimum	Higher
Master taping	$ 320	$1,000
Piano Tuner (including "standby")	$ 150	$ 240
Tape Editing	$ 0	$1,000
EQ and Transfer to CD Master	$ 400	$ 800
CD Reference Disc	(omit)	$ 150
Artwork and Typography	$ 500	$2,000
Sub-Total	*$1,370*	*$5,190*

Costs Dependent on How Many You Make, for 1,000 CD's

	Minimum	Higher
Paper and Printing	$ 500	$2,000
Discs, label printing, jewel boxes, insertion of booklets and inlay cards	$1,200	$1,400
Sub-Total	*$1,700*	*$3,400*
Total for 1,000 discs	$3,070	$8,590
Per Disc:	$ 3.07	$ 8.59
If 2,000 discs, Per Disc:	$ 2.39	$ 6.00

What final product will you make? For new releases, the Lp, with its well-known defects and little-known virtues, is all but dead. Cassettes can be made inexpensively, and can sound surprisingly good if made with care. The very best will startle you by how good they sound, but these are more costly than Compact Discs.

Digital Audio Tape (DAT) has somehow not been adopted as a consumer medium. The more recent media called Digital Compact Cassette (DCC) and Mini Disc (MD) were designed for convenience, not superior sound quality. Let's say you will make a CD.

Take your edited analog or digital master tape to a *mastering room*, or *transfer facility*, where it will be transferred to "CD-master" format for the plant that presses CD's. The mastering room is where any needed adjustment of tonal balance, lows to middles to highs (called "equalization," or just "EQ") is done. You need a good engineer to do a musically good job.

Let's say you and the mastering engineer need half an hour to decide on the right EQ. That might cost $125. The transfer of your analog tape to a CD master tape might cost $500 for a 60-minute CD. So far the total at the mastering room is $625; it might be less at some rooms, perhaps $400.

You might want a CD-REF (reference disc). This is a single custom CD made at the mastering room from your CD-master tape. It costs about $150, lets you know what your finished CD will sound like, and lets you verify the track entry points.

The mastering room sends the CD-master tape to the pressing plant, which charges you for making the needed glass CD master disc but waives the charge if you order a certain number of finished discs, say 1000. At the plant Performance Recordings® uses, the cost of finished CD's, including color printing of the labels (printed right on the discs), is about $1.18 apiece in "jewel cases" (those clear plastic boxes) with booklets and inlay cards inserted. (The inlay card is the printed sheet you see when you look at the back of the jewel case.)

As in everything, you almost never get more than you pay for. Often you get less.

There are companies that specialize in doing parts of the production work for you. My friend Rick Goldman's Compact Disc Service, in Glendale, California (818/241-9103), is one such. CDS will take your CD master tape and your label film (the photographic positive or negative from which the label will be printed) and produce 1000 finished discs in jewel boxes, without printed matter, for $1250. (There is no point in doing fewer, because it costs just about as much as 1000.) CDS will do the design and printing, too, for an additional charge.

Now you have your music on discs, but no booklets, inlay cards or labels. Actually, you have allowed for the cost of printing the labels but not for their design or typography; nor for design, typography, paper or printing of the booklet and inlay card.

There is a huge range of what you can pay for these things. You can do five-color printing on opaque "100 lb." paper, or two-color printing on 70 lb. paper that "shows through" what's printed on the other side. If you don't do your own design and typography on a computer, you can get exquisite or pedestrian design from a professional artist. And from a commercial "type house" you may get typography which lies evenly on the page and makes reading a pleasure, or which is so rough and jumpy it positively interferes with the communication of ideas.

The first alternative in each of these pairs is expensive; the second should be cheap, but sometimes isn't. As in everything, you almost never get more than you pay for. Often you get less. Even when you do get full value, reality sets limits. The late Paul Hunter, the artist I worked with for 16 years, said, "There are three things in commercial art: good, fast, cheap. You can have any two."

What will you do with your finished recordings? Sell them at your concerts! Commit the following to memory:

> *Good album reviews* do not *generate concert bookings*.
> *Concert bookings* do *sell recordings!*

A while after I made my first two recordings, I was playing at home for a group of friends. A soprano among them asked me if I thought recording had improved my playing. I said I thought so, and asked what difference she heard. She said my dynamic range was greater, which was just what I thought, too.

Dynamic range is one of the toughest things to preserve in any recording. Working to capture it made me aware of just how important it is to realistic reproduction, and this showed me its importance to the music. It became natural then to take more care over dynamics in my playing. The softs got softer, the louds got louder; and most important, the normal "tone of voice" went from *forte* to—well, *normal.*

Recording for release is thus not just a fascinating technical business. *Recording fosters artistic growth.*

Now what about those other uses for recording I mentioned at the beginning...?

Reader, we can tiptoe away from the descriptions I gave of how to use tape for dancing, singing, teaching, and the other things I have already told you about.

I hope that you have found our sessions rewarding, and that I have stimulated your interest rather than exhausted it.

Most of all, I hope you will actually try at least one of the activities I've suggested.

Remember: Practice performing and practice objectivity.

Objectivity
In the midst of involvement
Is our hardest task.

APPENDICES

Appendix A:

Recommended Listening and Reading

James Boyk, pianist and producer. *Pictures at an Exhibition*, by Modest Mussorgsky; original version for piano solo. Performance Recordings® pr7cd, pr7lp.

Unique analog-digital comparison. Available direct from Performance Recordings®, 310/475-8261. See Appendix B for further description.

James Boyk, Mark Fishman, Greg Jensen, Bruce Miller. *Demonstration of Stereo Microphone Technique*. Performance Recordings® pr6cd.

Unique; shows "stereo imaging" of 18 different miking techniques. Available direct from Performance Recordings®, 310/475-8261. See Appendix B for further description.

Barbara Conable, William Conable. *How To Learn the Alexander Technique*. Columbus, Ohio: Andover Road Press, 1992. ISBN 0-9622595-2-7.

Informal and friendly tone; packed with detailed information which is directly useful. Available through bookstores or direct from publisher at 1038 Harrison Avenue, Columbus, OH 43201.

Abby Whiteside. *Indispensables of Piano Playing*, 2nd ed. New York: Charles Scribner's Sons, 1961. ISBN 0684106531 and ISBN 0685045641.

Abby Whiteside. *Mastering the Chopin Etudes, & Other Essays*. Joseph Prostakoff and Sophia Rosoff, editors. New York: Charles Scribner's Sons, 1969. Out of print; no ISBN.

Appendix B:
James Boyk's Recordings and Articles

My boy, you may take it from me,
That of all the afflictions accursed,
With which a man's saddled
And hampered and addled,
A diffident nature's the worst.
—W. S. Gilbert, *Ruddigore*

Recording Credits

The author's credits are in **bold**.

All Performance Recordings® albums were recorded with the Blumlein technique described in Session 15; all but pr2 used ribbon microphones. Order direct from the Performance Recordings®, 310/475-8261. (Ordering information follows list of albums below.) Sheffield Lab albums all used ribbon microphones in a Blumlein arrangement.

Boyk plays Scarlatti and Beethoven *1978*
Scarlatti: Sonatas in C, K. 513; D Minor, K. 9;
G, K. 146. Beethoven: Sonata No. 32
in C Minor, Opus 111.
"Record to die for" (1992) —Stereophile
Performer / Producer / Album Notes
Performance Recordings® pr1 (Lp)
Out of print.

Boyk plays Schumann and Chopin *1980*
Schumann: "Kinderszenen," Opus 15.
Chopin: Fantasy in F Minor, Opus 49.
Performer / Producer / Album Notes
Performance Recordings® pr2 (Lp)
Virtuoso pianist…superb recording.
—Performing Arts
Out of print.

Boyk plays Prokofiev 1982
Sixth Piano Sonata.
Performer / Co-engineer / Producer / Album Notes
Performance Recordings® pr3 (Lp)
Boyk plays the very devil out of this sonata....
The piano sounds incredibly lifelike. —Stereo Review
Boyk's obsession with the purity of the recordings of his concerts has made him an authority on audio equipment. His Prokofiev is prized among audiophiles for being a rare example of a recorded piano sounding like one being played live. —Discover
Out of print. (Same performance available on CD. See below.)

Chicago Symphony Winds play Mozart & Grieg 1983
Mozart: Serenade No. 11 in E-flat Major, K. 375.
Grieg: Four Lyric Pieces, transcribed by Willard Elliot.
Microphone consultant
Sheffield LAB 22 (Lp), SLS 506 (CD)

Boyk plays Debussy, Stravinsky, Schoenberg, Ravel 1984
Debussy: "Reflections in the Water."
Stravinsky: Sonata {1924}.
Schoenberg: "Six Little Piano Pieces," Opus 19.
Ravel: Sonatina.
Performer / Co-engineer / Producer / Album Notes
Performance Recordings® pr4 (Lp)
Some of the most realistic and subtly nuanced piano sound I have heard on records in a long time....
The interpretation is worthy of the extraordinary care that has gone into making this record, and the result is something to cherish. —The Washington Post
Almost incredible fidelity...even more impressive is the integrity of the performance. —High Fidelity
(Same performances available on CD. See below.)

Boyk plays Beethoven 1985
Sonata in C Minor, Opus 13, "Pathétique";
Seven Bagatelles, Opus 33.
Performer / Co-engineer / Producer / Album Notes
Performance Recordings® pr5 (Lp)
(Same performances available on CD. See below.)

Kodo: Heartbeat Drummers of Japan 1985
Co-engineer (with Doug Sax)
Sheffield Lab CD-KODO (CD)
Superbly recorded...destined to become a classic.
—Hi-Fi News & Record Review

Firebird: Los Angeles Philharmonic/Leinsdorf *1985*
Debussy: Prélude to the Afternoon of
a Faun; Stravinsky: "Firebird" (1910)
Co-engineer (with Doug Sax)
Sheffield LAB 24 (Lp, CD)
The most convincing illusion of listening to a real, live orchestra from an excellent seat of any recording I have heard...an all-time great symphonic recording. —Stereophile

Demonstration of Stereo Microphone Technique *1990*
Producer / Co-engineer / Album Notes
Performance Recordings® pr6cd (CD)
Unique. Presents the imaging of 18 different stereo microphone techniques so that a layman can compare them easily.
The argument stopper. —Hi-Fi News & Record Review
Recommended. —National Public Radio Microphone Workshops
Highly recommended. —Audio

Boyk plays Mussorgsky *1991*
"Pictures at an Exhibition"
Performer / Co-engineer / Producer / Album Notes
Performance Recordings® pr7lp (Lp), pr7cd (CD)
World's only comparison of (a) pure digital, (b) digital-from-analog, and (c) pure analog recordings, made at the same time from the same microphones; (a) and (b) on the CD, (c) on the Lp.
Unbelievably precise, his musical concept extraordinarily definite. The piano sounds as a piano should. We rank his "Pictures at an Exhibition" among our reference recordings.
—HiFi Magazin (Hungary)
Perhaps the most distinguished interpretation I've ever heard... The feeling of "coming close to the vision of the composer" is strongly present through the whole piece.
—Musik & Ljudteknik (Sweden)

Boyk plays 20th Century Masters *1991*
(Complete programs of pr3 and pr4, above.)
Performer / Co-engineer / Producer / Album Notes
Performance Recordings® pr8cd

Boyk plays Beethoven *1991*
(Complete pr5 above plus Sonata in C-Sharp Minor, Op. 27, No. 2, "Moonlight.")
Performer / Co-engineer / Producer / Album Notes
Performance Recordings® pr9cd

Mayorga plays Chopin, Brahms, and Prokofiev 1992
(Chopin: 24 Préludes, Opus 28; Prélude, Op. 45; Prélude, Opus Posthumous. Brahms: Four Piano Pieces, Opus 119. Prokofiev: Three Pieces, Opus 96.)
Co-producer and Co-engineer (both with Doug Sax)
Sheffield Lab SLS 505 (CD)

How to Obtain the Recordings

Performance Recordings® albums are available direct from the maker at 2135 Holmby Avenue, Los Angeles, CA 90025-5915 USA. (Telephone: 310/475-8261, Fax: 310/470-9140, E-mail: boyk@caltech.edu.) Enclose US $20.00 per Lp or CD from North American addresses, $30.00 from anywhere else in the world. Price includes quick and safe shipping.

Sheffield Lab releases are available in stores or direct from the company at 213/466-3528.

Articles by James Boyk

In chronological order; includes talks for completeness. An asterisk (*) means of particular interest for readers of this book. References to EE/Mu 107 mean "Projects in Music & Science," James Boyk's course at California Institute of Technology, where he is Pianist in Residence, Lecturer in Music in Electrical Engineering, and Director of the Music Lab Auditioning Facility.

*"Keys to Success," *New West*, Oct. 24, 1977.
Nominally about how to choose a music teacher, but actually much broader. Called "the best statement of the philosophy of music education we've ever seen" by Yamaha International's Music Education Division, which distributed 11,000 reprints.

*"The Perfectly Complete, Completely Perfect, Thinking Person's Guide to Stereos," *New West*, Sept. 11, 1978.
Detailed advice on choosing stereo equipment by listening, guidance on choosing a store, etc. Finalist in the National Magazine Awards.

*"The Music Goes Round and Round and It Comes Out Here," *New West*, Jan. 1, 1979.
Introduction to recording and playback systems.

*"Oldies But Goodies," *New West*, Aug. 13, 1979.
Hi-fi equipment that has passed the test of time.

"The Cartridge Family," *New West*, Oct. 22, 1979.
Review of five phonograph cartridges.

"The Subjective vs. Objective Debate in Audio and the Current State of the Art," talk for Caltech's Communications Working Group. Repeated by invitation at Hewlett-Packard Co., Palo Alto, California, Mar. 24, 1981.

"Esthetic Perceptions and Technical Decisions in the Production of a Piano Recording," talk for Caltech's Communications Working Group.

*"The Ear of the Beholder," *New West*, July 14, 1980.
Beauty as a concrete aid in judging audio equipment.

*"The Music Lover's Quick But Accurate Guide to Stereo," *The Next Whole Earth Catalog*, 2nd edition, 1981.

*"The Music of Sound," guest editorial in *The Audio Amateur*, issue 5, 1982. Reprinted in *Hi-Fi News & Record Review* (England), Sept. 1985.
The sound itself is part of the meaning of the music.

"A Listening Test of Digital Audio," a double-blind test carried out as a project in EE/Mu 107. Talk for Caltech's Communications Working Group, Sept. 30, 1982. Repeated by invitation at Bolt Beranek Newman acoustical consultants, Canoga Park, Calif.

"Rules of the Game," *Hi-Fi News & Record Review* (England), Jan., 1983. Our perception of the New is shaped by mental categories established by our experience of the Old. How this affects our judgment of a variety of things including digital recording.

C. Minor, M. Todorovich, J. Boyk, G.P. Moore, "A Computer-Based Keyboard Monitor for Studying Timing Performance in Pianists," in *Timing and Time Perception*, J. Gibbon and L. Allen, eds. *Annals of New York Academy of Science* 1984; 423, 651.

Letter under heading "Controversy I: The Digital Wars," in *The Absolute Sound*, Summer, 1984.

"Comment," guest editorial, *Hi-Fi News & Record Review* (England), Dec., 1984. Remarks on philosophy of audio equipment evaluation.

"1960 vs. 1985 in Recording: A Quarter-Century of Degradation," Comparing analog recordings made with tube equipment in 1960 to digital recordings made with solid-state equipment in 1985. Talk for Caltech's Communications Working Group, April 11, 1985.

"Dynamic Inflection and the Beauty of Live Music," *Hi-Fi News & Record Review* (England), June, 1985.
Introducing a new term for equipment evaluations.

*"On Both Sides of the Microphone," *The Audio Amateur*, issue 1, 1986. An explicit account of how musical considerations affect technical decisions in recording.

"Piano Sound on Both Sides of the Microphone," talk for Patent Awards Banquet, GTE Laboratories, Waltham, Mass., June 25, 1987.

*"Backstage Info...on Microphones and Stereo," in booklet of Performance Recordings® album pr6cd, *Demonstration of Stereo Microphone Technique*. (See album list, above.)

"A Demonstration of Stereo Microphone Technique."
Discussing a technical demo recording made in EE/Mu 107, released as Performance Recordings® pr6cd. (See album list, above.) Talk for Caltech's Communications Working Group, Spring, 1990.

*"Audiences of the World, Arise!" *Harvard* magazine, Nov.-Dec., 1990. What's wrong with our concert halls, and why we continue to get bad halls. Reprinted as "When the Absolute Sound Isn't" in *The Absolute Sound*, issue 73, Sept.-Oct., 1991.

"An Analog-Digital Comparison in a Concert Recording." Comparing two recordings on the same CD: one made from digital master tape, the other from analog, both made at the same time from the same microphone feed. (Performance Recordings® pr7cd. See list above.) Talk for Caltech's Communications Working Group, Spring, 1991.

"10 feet 1¼ inches," *Los Angeles Times [Sunday] Magazine*, Oct. 1, 1995. Playing the world's biggest piano, the Fazioli F-308.

*"Essay: The Endangered Piano Technician," *Scientific American*, Dec., 1995. The disappearance of fine concert piano technicians threatens not just lovers of piano music but our whole musical culture.

Appendix C:

Technical Information on Selected Microphones

Additional information for Session 14, Audio Systems and Components. Ignore this appendix if you are not technically-minded!

Price	Make & Model	Output	Type	Pattern
$1,045	**Coles 4038**	1.8 mV	R	Figure-8
$ 895	*Sennheiser 441*	1.8 "	D	Supercardioid
$ 728	*AKG SE300B/CK94*	10. "	EC	Figure-8
$ 607	Shure SM81	5.6 "	C	Cardioid
$ 475	*Shure SM90A*	5. "	EC	Omni
$ 429	*AKG C-1000S*	6. "	EC	Cardioid/Hypercardioid
$ 399	**Beyer M-260**	1.2 "	R	Hypercardioid
$ 280	*Shure SM94*	3.2 "	EC	Cardioid
$ 228	*Shure 809*	4. "	EC	Omni
$ 211	Shure 849	2.8 "	EC	Cardioid
$ 208	*Audio-Technica 813A*	2.8 "	EC	Cardioid

Price: List price for one microphone is shown; however, some actually sell for less. The prices for the AKG SE300B/CK94 and the Shure models SM81 and SM90A include the necessary power supplies.

Make and Model: The two microphones in **boldface** are outstanding in sound quality, the Coles absolutely outstanding, the Beyer outstanding at its price. *Italic* means worthy at the price. I have not heard the other models. On paper, they look promising; but written descriptions and specifications cannot describe the sound. Listen before you buy.

Output: The higher the number, the more electrical pressure the microphone puts out for a given amount of sound pressure, and thus the easier it will be to make a tape with low background noise. *Any of these microphones can make a quiet tape with a separate microphone preamplifier, or with the Sony TC-D5M recorder.* The lower-output microphones will not make quiet tapes with the microphone preamplifiers built into some tape decks.

Technically, the figure given is the open-circuit voltage, in millivolts, for a sound level of 94 dB.

Type: R = Ribbon, D = Dynamic, C = Condenser, EC = Electret Condenser. These are names for various ways in which a microphone can convert sound into electricity. In my experience, the best sound has come from ribbon microphones.

Pattern: This describes directionality, indicating the microphone's sensitivity at various angles relative to the front. Superb microphones of each pattern exist. For your work in this book, one microphone is enough, and any pattern will be fine.

Two patterns are listed for the AKG C-1000S because it can be mechanically changed from one to the other. Of course it can be only one at a time.

The AKG SE300B/CK94 is the figure-8 version of a flexible microphone system that also offers a variety of other patterns by simply replacing the figure-8 "capsule" with the one of your choice. Thus, the SE300B/CK92 is an omni microphone.

For reasons beyond the scope of this book, the figure-8 pattern is best for recording in homes.

Excellent stereo (see Session 15) can be obtained with a pair of figure-8 microphones, or one figure-8 and one cardioid, or one figure-8 and one hypercardioid.

Appendix D:

Sources for Audio Equipment

Nationwide Dealers

Broadcast Supply West 800/426-8434
CAM Audio 800/527-3458
Mission Service Supply 800/352-7222
(Local dealers are in the Yellow Pages under "Stereo" or "Musical Instruments.")

Makers and Importers

Adcom 908/390-1130
AKG 818/894-8850
Audio-Technica 216/686-2600
Beyer 516/293-3200
B&W (Rotel of America) 508/664-3820
Coles (Wes Dooley) 818/798-9128
Harbeth (Musical Surroundings) 510/420-0379
KEF 508/429-3600
NAD 800/263-4641
NHT 800/648-9993
Rotel 800/370-3741
RTS 800/828-6107
Sennheiser 203/434-9190
Shure Bros. 800/257-4873
Sony Consumer Products 800/342-5721
Sony Professional Products 800/472-7669
Spendor (QS&D) 800/659-3711
Symetrix 800/288-8855

Reader's Update

TO HEAR OURSELVES AS OTHERS HEAR US

by James Boyk

Dear Reader,

To Hear Ourselves As Others Hear Us has been greeted with many generous comments since its publication last summer. Dr. John Zeigler, writing on the world-wide web's *Piano Education Page*, said, "Bravo to Mr. Boyk for this little gem of a book...that all piano teachers, students and parents should...read, and then read again. The insights and advice...are too many to skim through and too good to digest at only one exposure." *Engineering & Science* said, "The book teaches how to listen to oneself, and is richly illustrated with anecdotes from the author's own career. Boyk also includes a chapter on audio systems & components, giving readers the inside scoop from his many years testing recording equipment in his Caltech lab." And Dr. Bradford Gowen, writing in *Piano & Keyboard*, said, "All 'how to' books should be like this one: useful, cheerful, inspirational, thought-provoking, and so complete and clear as to be more than user-friendly—indeed, user-solicitous.... Boyk has the temperament to be both mechanical and artistic, and the wisdom to see both paths in the service of the same goal."

Hy Fujita

For this update, I recommend seven components out of 17 I've recently auditioned. I listened to each on albums for which I was performer, engineer or producer (see book, Appendix B); and on other outstanding albums, too. I also listened to the speakers directly from microphones in a nearby room, so I could judge the reproduction against the live sound; and I tested the recorders against live sound in my piano practice studio. Thus, these recommendations are not casual ones.

Loudspeakers. Of nine models I heard, I recommend the six below. The first two sounded better to me than the others. The remaining four are in alphabetical order. Remember that all six are recommended. Listings give make, model and price per pair, then dimensions in inches (HWD). "Shielded" means the manufacturer says the unit may be used near TVs or computer monitors. Call the number given to find your nearest dealer.

Mordaunt-Short MS30i, $529, 16.8x9.8x11, not shielded, 1-800-663-9352.

NHT SuperOne, $350, 11.65x7.25x8.5, shielded, 1-800-648-9993.

NHT donated a pair of model 3.3 to the Caltech music lab, which I direct; and sold us 1.1s at "academic price"; but I don't think this has affected my judgment of the SuperOne.

B&W DM601, $400, 14x8x9.625, not shielded, 1-800-370-3740.

Celestion Impact 20, $500, 18.5x10.3x11.4, shielded, 1-800-356-9470.

Polk RT5, $329.90, 14.5x8.5x9.75, shielded, 1-800-992-2520.

Polk RT7, $449.90, 19x9.5x11.5, shielded, 1-800-992-2520.

Amplifiers. I listened to models at list prices of $199, $299 and $399, but can't recommend any of them. I haven't heard any cheap units with convincing sound since the long-discontinued NAD 3020, which sold in the 1980's for under $200. *Please, NAD, bring back the 3020!*

Cassette recorders. I listened to models that list for $499, $599 and $695, but can't recommend any of them. There's no reason why a good machine for our purposes couldn't sell for $400 or less, yet of the machines I've tried, the cheapest one I can recommend is still the excellent *Sony TCD-D5* (page 47), which sells at discount for $600.

Digital audio tape (DAT) recorders. I evaluated two portables. One was cute but not recommendable. The other is the *Tascam DA-P1* ($1899 list, $1299 discount; 1-213-726-0303, ext. 834 for nearest dealer), which feels solid and has quiet mike preamps, so you may use any microphone you choose. However, I found its sound quality somewhat "generic": it seemed to make my Steinway sound like a synthesized piano. I could hear my music-making more clearly with the Sony TCD-D5 (see above). The first sample of the DA-P1 was defective; the second worked fine. According to the manual, the FCC Class A rating means that "Operation of this equipment in a residential area is likely to cause harmful interference...."

MIDI. A few readers have asked about recording themselves with MIDI devices. Attractive as the idea is, I can't recommend it. MIDI works by moment-to-moment sampling; but unlike digital audio, which samples your *sound*, MIDI samples your *actions*. (Play G medium-loud, hold the G while playing F softly, put the pedal down while holding both notes, let go of the F; and so on.) Also unlike digital audio, which samples the sound very frequently, MIDI samples your actions quite infrequently, and thus cannot reproduce your timing precisely. If it's hard to dance to a MIDI playback, you can't be sure whether the problem is with you or the MIDI recording. Moreover, the MIDI samples don't correctly capture how loudly each note was played; and overall, MIDI seems to narrow your dynamic range. Since timing and dynamics are crucial to a performance, in my opinion you can't trust MIDI (as it currently exists) to reproduce your performance accurately.

To audio manufacturers. Every student of every instrument needs a musically accurate recording and playback system. So does every teacher, while every music school needs several. Think how many potential sales these groups represent! Make the next great cheap amplifier! Make a cheap recorder with quiet preamps & stable speed! Make a "musician's recorder," with special features for practicing and teaching! Call me for advice! (818 395-4590, <boyk@caltech.edu>)

Yours very truly,

James Boyk

For further information and catalogs, contact:

MMB Music, Inc.
Contemporary Arts Building
3526 Washington Avenue
Saint Louis, MO 63103-1019

Phone: 314 531-9635, 800 543-3771 (USA/Canada)
Fax: 314 531-8384
E-mail: mmbmusic@mmbmusic.com
Web site: http://www.mmbmusic.com